D0025408

$\mathcal{D}$-MODULES AND SPHERICAL REPRESENTATIONS

BY

FREDERIC V. BIEN

MATHEMATICAL NOTES
PRINCETON UNIVERSITY PRESS

$\mathcal{D}$-modules and Spherical Representations

$\mathcal{D}$-modules and Spherical Representations

by

Frédéric V. Bien

Mathematical Notes 39

PRINCETON UNIVERSITY PRESS

PRINCETON, NEW JERSEY

1990

Princeton University Press books are printed on acid-free paper, and meet the guidelines for permanence and durability of the Committee on Production Guidelines for Book Longevity of the Council on Library Resources

Printed in the United States of America
by Princeton University Press, 41 William Street
Princeton, New Jersey

Library of Congress Cataloging-in-Publication Data

Bien, Frédéric V., 1961–
D-modules and spherical representations / by Frédéric V. Bien.
p. cm. — (Mathematical notes ; 39)
Includes bibliographical references and index.
ISBN 0-691-02517-7 (acid-free paper)
1. Differentiable manifolds. 2. Representations of groups.
3. Lie groups. I. Title. II. Series: Mathematical notes
(Princeton University Press) ; 39.
QA614.3.B5 1990 514'.74—dc20 90-8776

A mes parents Nadine et Hubert
et à ma muse Tesa.

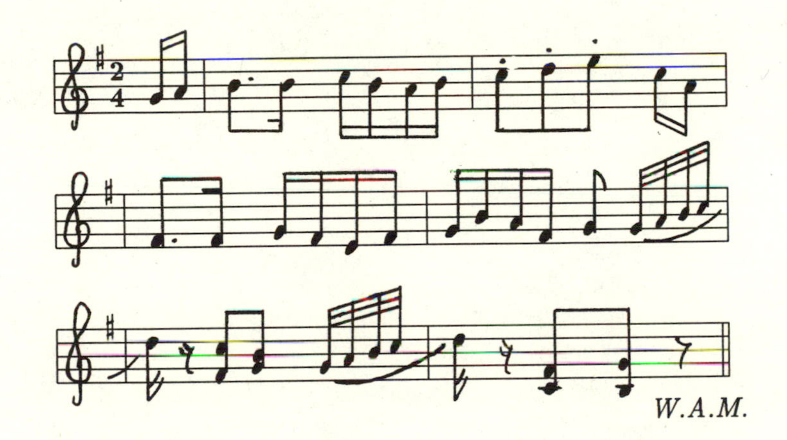

Acknowledgements

The main part of this monograph consists of my Ph.D. thesis [Bien2]. I wish to express my sincere gratitude to my advisor Joseph Bernstein, for generously sharing with me his understanding of mathematics, and stimulating my work with inspiring conversations. I am also very grateful to David Vogan for helpful discussions, and for examining carefully early versions of the "SUNscript". Without him, this book would have appeared in a poor state of accuracy. He provided me with a list of errors and detailed comments which improved greatly the presentation. I apologize for the imperfections that undoubtedly remain.

Special thanks go to Mogens Flensted-Jensen, Pierre-Yves Gaillard, Bertram Kostant, and Toshio Oshima for valuable pieces of advice and insightful comments. It is a pleasure to acknowledge the support of the following institutions during the preparation of this monograph: the Belgian American Educational Foundation, the Massachusetts Institute of Technology, the Alfred P. Sloan Foundation, the National Science Foundation, the Institute for Advanced Study, and Princeton University.

Finally, I thank Becci Davies who typed the first version of this book using LaTeX. The many succeeding versions were made by the author.

Contents

Introduction **1**

I Localization Theory **9**

I.1 Basic notions on $\mathcal{D}$-modules . . . 9
I.2 $\mathcal{D}$-modules with a group action . . . 10
I.3 Parabolic subgroups and flag spaces . . . 12
I.4 Differential operators on flag spaces . . . 14
I.5 Global sections of $\mathcal{D}_\lambda$ . . . 15
I.6 $\mathcal{D}$-affine varieties . . . 19
I.7 Holonomic modules and Harish-Chandra modules . . . 24

II Spherical $\mathcal{D}$-modules **29**

II.1 K-orbits in a flag space. . . . 29
II.2 $(\mathcal{D}, K)$-modules with a K-fixed vector . . . 33
II.3 Relations with the analytic theory. . . . 34
II.4 Going from $(\mathcal{D}, K)$ modules to $(\mathcal{D}, H)$ modules . . . 39
II.5 H-spherical $(\mathcal{D}, K)$-modules. . . . 45
II.6 $\mathcal{D}$-modules and hyperfunctions. . . . 52
II.7 Closed orbits and discrete series. . . . 57
II.8 An algebraic Poisson transform. . . . 63

III Microlocalization and Irreducibility **68**
III.1 Microlocal Differential Operators 68
III.2 Microlocalization of a non-commutative ring. 69
III.3 Microlocalization of $\mathcal{D}_\lambda$ 71
III.4 Microlocalization of U_χ 73
III.5 Microlocal study of the moment map. 74
III.6 Associated variety of a submodule of $\Gamma\mathcal{M}$ 76
III.7 Irreducibility criterion 78
III.8 Decomposable modules. 80

IV Singularities and Multiplicities **84**
IV.1 Normality 84
IV.2 Unibranchness 86
IV.3 Invariant differential operators on a reductive symmetric space 89
IV.4 Multiplicity one within an eigenspace of $\mathbf{D}(G/H)$ 96
IV.5 Intertwining maps between distinct eigenspaces of $\mathbf{D}(G/H)$ . . 108
IV.6 Multiplicities of principal series 116
IV.7 τ-invariant of discrete series on exceptional symmetric spaces . 118

Bibliography **125**

Index **130**

Introduction

The goal of this work is to study the representations of reductive Lie groups which occur in the space of smooth functions on an indefinite symmetric space. The representations realized by square integrable functions were constructed by Flensted-Jensen, Oshima and Matsuki. We prove that the discrete series has multiplicity one, except possibly in a few exceptional cases, and we present a cohomological formula for the multiplicities of standard representations. We study these representations by Beilinson and Bernstein's theory of differential operators on complex flag manifolds, and we present a summary of this theory. We also find a canonical map between certain $\mathcal{D}$-modules and the sheaf of hyperfunctions along a real flag manifold. Combined with previous work of Helgason and Flensted-Jensen, this exhibits a natural intertwining operator between these $\mathcal{D}$-modules and functions on symmetric spaces. By microlocalization, we also make a finer study of the moment map, and decide the irreducibility of the global sections of $\mathcal{D}$-modules considered as representations of Lie algebras.

More precisely, let G be a complex connected reductive linear algebraic group, and let G_o be a real form of G. Consider the fixed point subgroup H of an involution σ of G and its corresponding real form H_o. Then G_o/H_o is a symmetric space. The problem is to find which representations of G_o can be imbedded in $C^\infty(G_o/H_o)$, which ones can be imbedded in $L^2(G_o/H_o)$ and when they can be so imbedded, in how many different ways.

To formulate the answer, one should first understand that the building blocks of $C^\infty(G_o/H_o)$ are generally not the irreducible representations of G_o. The standard representations (defined in I.7) are better designed for this purpose but not all of them will appear. A smooth representation V of G_o which does appear, is called H_o-*spherical* and is characterized by: $\mathrm{Hom}_{H_o}(V, \mathbf{C}) \neq 0$; in other words, its continuous dual V^* contains an H_o-fixed functional v_0. If we want V to be imbedded into $C^\infty(G_o/H_o)$, then v_0 must generate V^*. The duality involved is important because in general V itself does not contain any H_o-invariant vector.

Choose a σ-stable maximal compact subgroup K_o of G_o, with complexification K. Let $\mathbf{g}$, $\mathbf{h}$ and $\mathbf{k}$ denote the Lie algebras of G, H and K respectively. Let V^0 be the Harish-Chandra module of V: it is the $(\mathbf{g}, K)$-module made of all K-finite vectors in V. A few years ago, [Casselman, p. 44] and [Wallach,

p. 320] discovered an 'automatic continuity' theorem for linear functionals invariant under a maximal nilpotent subalgebra. Their method can be extended to our situation, to show that a linear functional on V^0 invariant by $(\mathbf{h}, K \cap H)$ has automatically a continuous extension to V. The proof has been spelled out in [van den Ban–Delorme]. Therefore we can replace the H-spherical condition by $\mathrm{Hom}_{(\mathbf{h}, K\cap H)}(V^0, \mathbf{C}) \neq 0$.

Using the inductive version of Zuckerman's functor (defined in II.4), this is equivalent to $\mathrm{Hom}_H(L_K^H V^0, \mathbf{C}) \neq 0$, that is $(L_K^H V^0)^H \neq 0$ since H acts semisimply on this space.

We pass to the theory of $\mathcal{D}$-modules. Helgason's isomorphism for Riemannian symmetric spaces shows that the spherical representations have a particular shape: they can be realized on a partial flag variety. The Iwasawa decomposition associates to G and H a complex flag variety X, and to the datum of G_o, we can associate a real submanifold X^r imbedded in X. To fix the ideas, consider a real form G^r of G for which $H^r := H \cap G^r$ is a maximal compact subgroup. Then, the symmetric space G^r/H^r is *Riemannian*, (this justifies our notation), and it is called the dual of G_o/H_o. Let P^r be a minimal parabolic subgroup of G^r with complexification P. We put $X \simeq G/P$ and $X^r \simeq G^r/P^r$; X^r is called the dual flag space of G_o/H_o. Helgason's isomorphism shows that the eigenspace representations of G^r on G^r/H^r are equivalent to the representations in sections of line bundle on X^r. It may seem to the reader that we are taking a round about approach, beacause after all G_o has also a real flag space X_o, namely its only closed orbit in X. But the relation betwen eigenspace representations on G_o/H_o and sections on line bundles on X_o is not as well-understood as in the Riemannian case.

Let $T_P = P/(P,P)$ be the Cartan factor of P. Up to covering, it is isomorphic to the center of the Levi factor of P. To every $\lambda \in \mathbf{t}_p^*$, one can associate a G-sheaf $\mathcal{D}_\lambda$ of twisted differential operators on X. To define the notion of H-spherical $(\mathcal{D}_\lambda, K)$-modules, we will use the functor $\mathcal{L}_K^H$ introduced by Bernstein: it maps $(\mathcal{D}_\lambda, K)$-modules to $(\mathcal{D}_\lambda, H)$-modules in a fashion analogous to the inductive Zuckerman's functor. Assume henceforth that $\lambda \in \mathbf{t}_p^*$ is dominant, so that the global section functor is exact. We say that a $(\mathcal{D}_\lambda, K)$-module is *H-spherical* if

$$\Gamma(X, \mathcal{L}_K^H \mathcal{M})^H \neq 0$$

The standard $(\mathcal{D}_\lambda, K)$-modules on X are constructed as follows. Take a K-orbit Y affinely embedded in X with a K-homogeneous line bundle $\mathcal{O}_Y(\lambda)$

determined by the one-dimensional representation $\mathbf{C}(\lambda-\rho_p,\tau)$ of the isotropy pair $(\mathbf{p}_y, K_y)$ for any $y \in Y$. Let i_* be the direct image in the category of $\mathcal{D}_\lambda$-modules. Then $\mathcal{M}(Y,\lambda) = i_*\mathcal{O}_Y(\lambda)$ is a standard $(\mathcal{D}_\lambda, K)$-module. Now let X^0 denote the open H-orbit in X and let L be the Levi factor of P.

Theorem 1: (II.5.3)[1] *Let $\lambda + \rho_L$ be B-dominant and regular. $\mathcal{M}(Y,\lambda)$ is H-spherical if and only if $Y \cap X^0 \neq \emptyset$ and $(\mathbf{h}_y, K \cap H_y)$ acts trivially on the fiber $\mathcal{O}_Y(\lambda)_y$, $y \in Y \cap X^0$.*

Let $L^2_K(G_o/H_o)$ be the space of K_o-finite vectors in $L^2(G_o/H_o)$. A $(\mathcal{D}_\lambda, K)$ -module whose global sections can be imbedded as a discrete summand of $L^2_K(G_o/H_o)$ is called H-square integrable. Using [Oshima – Matsuki] L^2-estimates, we can simplify their result on the discrete series of G_o/H_o. Suppose G is semisimple.

Theorem 2: (II.7.5) *$\mathcal{M}(Y,\lambda)$ is H-square integrable if and only if $\mathcal{M}(Y,\lambda)$ is H-spherical, Y is a closed K-orbit, $\lambda \in \mathbf{t}^*_p$ is dominant regular and* $\operatorname{rank} G/H = \operatorname{rank} K/H \cap K$.

Let $\mathbf{t}$ be the canonical Cartan space of G, i.e. the Cartan factor of a Borel subalgebra of G. If $\lambda + \rho_\ell \in \mathbf{t}^*$ is dominant, then $\Gamma(X, \mathcal{M}(Y,\tau))$ is an irreducible $(\mathbf{g}, K)$-module. But in general there is a small strip of dominant $\lambda \in \mathbf{t}^*_p$ for which $\lambda + \rho_\ell$ is not dominant. [Vogan 5] has checked by coherent continuation that $\Gamma(X, \mathcal{M}(Y,\tau))$ remains irreducible in this strip. We have proved independently the main ingredient of his result (theorem III.7.2), and we obtain an interesting consequence of irreducibility, which in essence motivated this whole work.

Theorem 3: (II.7.6, IV.3.6, IV.5.1&1') *The discrete series in $L^2(G_o/H_o)$ is multiplicity free except possibly for a few exceptions that are explained below; in these exceptional cases, the multiplicities can be at most two if* $\mathbf{g}$ *is simple.*

The exceptions to theorem 3 can occur only for reductive symmetric spaces containing factors of type $E6/(\mathbf{so}(10)\oplus\mathbf{C})$, E_6/F_4, $E_7/(E_6\oplus\mathbf{C})$, $E_8/(E_7\oplus\mathbf{sl}(2))$, and on these spaces, the representations that could have multiplicity two, must have singular infinitesimal characters in range that is listed in section IV.7. Thus, the discrete series of all classical symmetric spaces are multiplicity free.

[1] The numbers between parentheses refer to the places where the results are proved in this book.

Theorem 3 implies that all connected compact symmetric spaces are multiplicity free, a fact that goes back to work of E. Cartan and S. Helgason. Note also that for Riemannian symmetric spaces, the statement of theorem 3 is vacuous, since there is no discrete series. However, one knows in this case, that the continuous series is multiplicity free. These two properties were formalized into the notion Gelfand pairs, *cf.* also work of B. Kostant. As we shall see below, if H_o is not compact, then (G_o, H_o) is very rarely a Gelfand pair.

Let χ denote an eigencharacter of the algebra $\mathbf{D}(G/H)$of invariant differential operators on the symmetric space G_o/H_o and let $C^\infty(G_o/H_o;\chi)$ be the corresponding eigenspace. Note that one should distinguish between the multiplicity of an irreducible representation V of G as subquotient of $C^\infty(G_o/H_o,\chi)$ and as submodule. The former can be quite big although finite. The latter is $\dim(V^*)^H$. Theorem 3 means that this dimension is at most 1, for $V = \Gamma(X, \mathcal{M}(Y,\lambda))$ as above, except possibly for a few cases.

In fact the determination of these multiplicities for submodules motivates this whole work. We are able to interpret $(V^*)^H$ as the cohomology space of a smooth algebraic variety with values in a local system. This formula holds for all standard modules obtained from affinely imbedded K-orbits. If V is reducible, it only determines the dimension of the space of morphisms: $\mathrm{Hom}_{G_o}(V, C^\infty(G_o/H_o))$. For any x in the open H-orbit X^0, let M be the isotropy group of x in H. The invertible sheaf $\mathcal{O}_Y(\lambda)$ gives rise to a $K\cap H$-homogeneous local system $\mathbf{C}(\lambda,\tau)$ on $Y\cap X^0$.

Theorem 4: (II.5.5)*Suppose that $\lambda+\rho_L$ is B-dominant and regular. Then:*

$$\Gamma(X, \mathcal{L}_K^H \mathcal{M}(Y,\tau))^H \simeq H^s(Y\cap X^0, \mathbf{C}(\lambda,\tau))^{K\cap H}$$

where $s = \dim Y + \dim M - \dim K\cap H$.

This formula easily gives the uniqueness of the H_o-invariant functional in many nontrivial cases. It also shows that the even principal series for $SL_2(R)$ admits generically two $SO(1,1)$-invariant functionals.

It is worth noting that exponential symmetric spaces, in particular symmetric spaces associated to nilpotent Lie groups, satisfy the multiplicity one property in their whole L^2 spectrum. This was proved by [Benoist] using a method similar to the one used for uniqueness of Whittaker functionals. This method does not work for reductive groups. In fact, if it did work, it would show that the whole L^2-spectrum is multiplicity free, which is not true as mentioned above. Nevertheless, it would be interesting to prove that for a symmetric space associated to a *general* Lie group, the discrete part of the L^2-spectrum is multiplicity free.

Before describing the content of the chapters, we would like to refer the neophytes to two expository books where they can find essentially all the prerequisites needed to understand the techniques used here: [Borel] for the algebraic theory of $\mathcal{D}$-modules, and [Schlichtkrull] for familiarity with symmetric spaces and the basic properties of hyperfunctions.

In chapter I, we summarize [Beilinson–Bernstein] theory of $\mathcal{D}$-modules on flag spaces and its relations to the representations of $\mathbf{g}$. We take a general viewpoint by working over the variety X of parabolic subgroups of G conjugate to a given one, say P. The vanishing theorem for $\mathcal{D}_\lambda$-modules is true as long as λ is dominant in $\mathbf{t}_P^*$. But $\Gamma(X, \mathcal{D}_\lambda)$ need not be generated by the enveloping algebra $\mathcal{U}(\mathbf{g})$ for all $\lambda \in \mathbf{t}_p^*$, as it is the case for the variety of Borel subgroups. This is due to the fact that the closure of the Richardson orbit $C_P \subset \mathbf{g}^*$ of P need not be a normal variety. However, if $\lambda + \rho_\ell \in \mathbf{t}^*$ is dominant, then $\mathcal{U}(\mathbf{g})$ generates $\Gamma(X, \mathcal{D}_\lambda)$. If in addition $\lambda + \rho_\ell$ is regular, the category of $(\mathcal{D}_\lambda, K)$-modules on X is equivalent to a subcategory of the category of all $(\mathbf{g}, K)$-modules with infinitesimal character $\lambda + \rho_\ell$ and associated variety contained in $\overline{C_P}$. To study $\Gamma(X, \mathcal{D}_\lambda)$, we generalize a method of Miličić. The non-vanishing of the global sections of a $(\mathcal{D}_\lambda, K)$-module depends on the Borel-Weil theorem. To include entirely this classical result in the theory of $\mathcal{D}$-modules, one should consider modules over a sheaf of matrix differential operators. Thus, one could generalize the theory much further than we have done here.

In chapter II, after recalling some properties of the K-orbits in X, we study the case $K = H$ of Riemannian symmetric space. We classify the K-spherical $(\mathcal{D}_\lambda, K)$-modules: they all come from trivial line bundles on the open K-orbit. In particular, there is no square integrable module and they all have a unique K-invariant vector. These results have been known since [Kostant]. In section II.3 we explain what is the foundation of our approach for the reader with an analytic background. Next we describe Bernstein's definition of Zuckerman's functor for $\mathcal{D}$-modules. At this point we are ready to study the H-spherical $(\mathcal{D}_\lambda, K)$-modules. The idea is again that they are strongly related to the open H-orbit and to homogeneous line bundles which are trivial for $(\mathbf{h}, K \cap H)$. We first prove theorem 1. Then we prove theorem 4 by computing the number of $(\mathcal{O}_X, H)$-morphisms from $\mathcal{O}_X$ into $\mathcal{L}_K^H \mathcal{M}(Y, \tau)$ for a standard $(\mathcal{D}_\lambda, K)$-module $\mathcal{M}(Y, \tau)$, in terms of the cohomology of $Y \cap X^0$. A better understanding of this formula would give a purely algebraic classification, but at the present time we are obliged to use as an intermediate step the real flag variety X^r of G^r which is imbedded in the middle of X^0. In some sense this is fortunate because we can construct a map between K-orbits Y in X such that $Y \cap X^r \neq 0$ and Norm(G^r, K)-orbits in X^r. Then we answer a question raised by Flensted-Jensen in a lecture at Utrecht in August 1985: what is the relation between the square-integrable standard H-spherical $(\mathcal{D}_\lambda, K)$-modules and the space of hyperfunctions sections of a line bundle on X^r? It turns out that the first can be imbedded into the second, thanks to the existence of a canonical morphism of functors $f^* \to f^! \otimes \omega[-rdf]$ which exists for any map f. Let us remark that tracing back through Helgason and Flensted-Jensen isomorphisms, this map gives the unique imbedding of square integrable standard H-spherical $(\mathcal{D}_\lambda, K)$ module into $L^2(G_o/H_o)$. This shows that the $\mathcal{D}$-module realization of a Harish-Chandra module is in fact quite natural to study harmonic analysis on symmetric spaces. Note that the spaces G_o/H_o^0 can be handled in a similar manner, where H_o^0 is the identity component of H.

Then we focus on closed K-orbits in the equal rank case, and using an L^2-estimate of Oshima and Matsuki, we prove theorem 2. Next we show that the discrete series of G_o/H_o has multiplicity one in the generic range. Finally, we define an algebraic Poisson transform from $(\mathcal{D}_\lambda, K)$-module on G/P to $(\mathcal{D}_{\chi_\lambda}, K)$-modules on G/H. [De Concini and Procesi] have constructed a very nice compactification of G/H which exhibits G/P as a piece of the boundary of G/H. I would expect the nearby cycle functor to be the inverse of the Poisson transform.

In chapter III, we solidify the bridge between $\mathbf{g}$-modules and $\mathcal{D}_\lambda$-modules on an arbitrary flag variety X. Chapter III is independent of chapter II. Its motivation is to understand when does the global section functor preserves irreducibility. Let U_χ be the quotient of $U(\mathbf{g})$ by the ideal determined by $\lambda \in \mathbf{t}_p^*$. The image of the moment map $\pi : T^*X \to \mathbf{g}^*$ is the closure of the Richardson orbit $C_P \subset \mathbf{g}^*$ of P. Following a method introduced by Gabber and [Ginzburg], we define the formal microlocalizations $\mathcal{E}_\lambda, \mathcal{U}_\chi$ of the algebras $\mathcal{D}_\lambda$, $\mathcal{U}_\chi$. $\mathcal{E}_\lambda$ is a G-sheaf of twisted formal microdifferential operators on T^*X, and $\mathcal{U}_\chi$ is a G-sheaf supported on $\overline{C_P}$. The microlocalization is an exact and faithful functor which sends a $\mathcal{D}$-module $\mathcal{M}$ to an $\mathcal{E}$-module whose support is the characteristic variety char $\mathcal{M}$ of $\mathcal{M}$. The functors Γ and Δ of global sections and localizations between $\mathcal{D}_\lambda$- and U_χ-modules become the functors π_* and π^* of direct and inverse images between $\mathcal{E}_\lambda$ and $\mathcal{U}_\chi$-modules. One result of this approach is the following.

Theorem 5:(III.6.4) *Suppose $\mathcal{M}$ is an irreducible $\mathcal{D}_\lambda$-module on X and λ is dominant in $\mathbf{t}_\mathbf{p}^*$. Then all the irreducible $U(\mathbf{g})$-submodules of $\Gamma(X, \mathcal{M})$ have the same associated variety.*

The main result of this chapter is a criterion for irreducibility.

Theorem 6:(III.7.2)*Suppose the moment map π is birational. Let $\mathcal{M}$ be an irreducible $\mathcal{D}_\lambda$-module with λ dominant in $\mathbf{t}_\mathbf{p}^*$. If π is an isomorphism in a neighborhood in $\overline{C_P}$ of the generic point of the characteristic variety of $\mathcal{M}$, then $\Gamma(X, \mathcal{M})$ is $U(\mathbf{g})$-irreducible or zero.*

In practice, the first hypothesis on π is easily checked using (IV.1.1), and the second hypothesis amounts to checking that $\pi(\text{Char}\mathcal{M})$ contains some normal points of $\overline{C_P}$. Each of these results has its counterpart for the K-equivariant situation.

In chapter IV, we come back to the original question of determining the multiplicities. First we study the singularities of the image $\overline{C_p}$ of π. For the classical groups [Kraft and Procesi] have determined when $\overline{C_p}$ is a normal variety. This information yields theorem 3 for the classical groups and G_2.

The microlocal approach and the study of examples suggests that the unibranchness of an open dense subset of $\pi(\text{char}\,\mathcal{M})$ should suffice to imply the irreduciblity of $\Gamma(X, \mathcal{M})$ over $(\mathbf{g}, K)$ when $\mathcal{M}$ is $(\mathcal{D}_\lambda, K)$-irreducible and π birational. Generalizing a result of [Borho-MacPherson], we find a formula which detects the unibranchness of any point $x \in \overline{C_p}$ in terms of the multiplicity of the special Weyl group representation attached to C_p in a cohomology space of the Springer fiber of x.

To prove theorem 3 for the exceptional groups, we must proceed in two steps. First, we establish multiplicity one within one eigenspace of $\mathbf{D}(G/H)$: the algebra of invariant differential operators on G_o/H_o. Next, we prove a new result which asserts that an endomorphism of $L^2(G_o/H_o)$ commuting with G_o and having a regular kernel, must also commute with $\mathbf{D}(G/H)$, (IV.5.2). Here, regular kernel means that the kernel is determined by its restriction to the regular set. Unfortunately, this result does not imply that a discrete series representation cannot appear in two different eigenspaces of $\mathbf{D}(G/H)$, since there exist examples of H_o-invariant eigendistribution vectors corresponding to discrete series representations which are supported on the singular set.

In section IV.6, we determine the multiplicities of the principal series representations in $L^2(G_o/H_o)$ for generic values of the parameter λ. Consider a standard H-spherical $(\mathcal{D}_\lambda, K)$-module $\mathcal{M}(Y,\lambda)$ associated to the open K-orbit Y in X. Then $M(Y,\lambda) := \Gamma(X, \mathcal{M}(Y,\lambda))$ is an irreducible $(\mathfrak{g}, K)$-module for λ generic. Let $K^\tau = K \cap G^\tau$. Then $Y^\tau := Y \cap X^\tau$ breaks into a finite number – say n – of K^τ-orbits.

Theorem 7: (IV.6.1) *The multiplicity of $M(Y,\lambda)$ in $C^\infty(G_o/H_o;\chi_\lambda)$ is equal to n.*

This result explains the origin of the multiplicities for symmetric spaces. In particular, the multiplicity one theorem for the discrete series can be understood by the fact that if Y is a closed K-orbit such that $Y \cap X^0$ is not empty, then $Y \cap X^\tau$ consists of a single K^τ-orbit.

The last section contains the computation of the τ-invariant of singular discrete series representations on exceptional symmetric spaces. Even this deep method turns out to be insufficient to determine the multiplicities of exceptional cases in theorem 3.

Thus, the story expounded in this book, ends in an interesting suspense. We draw many conclusions, and unveil a moral. However, our heroes may still dream of surprising adventures. Hence, let me keenly encourage the young readers – and the young at heart – to finish the proof of the multiplicity one theorem, and more generally to determine all the multiplicities by a computationally practical formula.

I Localization Theory

I.1 Basic notions on $\mathcal{D}$-modules

Let X be a smooth complex algebraic manifold. Let $\mathcal{O} = \mathcal{O}_X$ be the sheaf of regular functions on X and $\mathcal{D} = \mathcal{D}_X$ be the sheaf of differential operators on X . A $\mathcal{D}$-module $\mathcal{M}$ is a sheaf on X which is quasi-coherent as an $\mathcal{O}$-module and which has a structure of module over $\mathcal{D}$. Let $\mathcal{D}$-mod denote the category of left $\mathcal{D}$-modules. We shall deal with some sheaves of rings slightly more general than $\mathcal{D}$. Consider the category of pairs $(\mathcal{A}, i_{\mathcal{A}})$ where $\mathcal{A}$ is a sheaf on X of $\mathbf{C}$-algebras and $i_{\mathcal{A}} : \mathcal{O} \to \mathcal{A}$ is a morphism of $\mathbf{C}$-algebras. The pair $(\mathcal{D}, i)$ where $i : \mathcal{O} \to \mathcal{D}$ is the natural inclusion will be called the standard pair.

1.1 Definition: A sheaf of twisted differential operators on X (tdo for short) is any pair $(\mathcal{A}, i_{\mathcal{A}})$ which is locally isomorphic to the standard pair.

1.2 Lemma: *The group of automorphisms of $(i : \mathcal{O} \to \mathcal{D})$ is naturally isomorphic to $\mathcal{Z}^1(X)$ the group of closed 1-forms on X . The set of isomorphism classes of tdo's on X is in bijection with the Cech cohomology space $H^1(X, \mathcal{Z}^1)$.*

In particular tdo's form a linear space; for a proper variety X , $H^1(X, \mathcal{Z}^1)$ is the $\mathbf{C}$-subspace of $H^2_{dR}(X)$ generated by algebraic cycles.

1.3 Example: Let $\mathcal{L}$ be an invertible sheaf of $\mathcal{O}$-modules on X ,then the sheaf $\mathcal{D}(L)$ of differential operators acting on the local sections of $\mathcal{L}$ is a tdo. Let $\mathcal{L}$ correspond to the element c in the Picard group $H^1(X, \mathcal{O}^*)$, then $\mathcal{D}(L)$ corresponds to the logarithmic differential of c, i.e. dc/c in $H^1(X, \mathcal{Z}^1)$.

1.4 Example: If $X = \mathbf{P}^n$ is the projective n-space, then $H^1(X, \mathcal{O}^*) = \mathbf{Z}$, hence, the invertible sheaves on $\mathbf{P}^n$ form a lattice. But $H^1(\mathbf{P}^n, \mathcal{Z}^1) = \mathbf{C}$: the sheaves of twisted differential operators on $\mathbf{P}^n$ form a vector space.

1.5 Remark: If $\mathcal{V}$ is a locally free sheaf of $\mathcal{O}$-modules on X of rank bigger than one, then the sheaf of rings of differential operators of $\mathcal{V}$ into itself is not a tdo. It consists of matrix differential operators. This sheaf will be useful in chapter IV.4.

Let $f : X \to Y$ be a morphism of smooth complex algebraic varieties.

Let $\mathcal{O}_Y, \mathcal{O}_X$ be the sheaves of regular functions on Y and X. We will denote by $f^\circ, f_\circ$ the inverse and direct images functors between the categories of $\mathcal{O}$-modules. There is an induced map $f^* : H^1(Y, \mathcal{Z}^1) \to H^1(X, \mathcal{Z}^1)$. Starting from a tdo $(\mathcal{D}_{Y,\lambda}, i)$ on Y corresponding to $\omega \in H^1(Y, \mathcal{Z}^1)$, we can construct the tdo. $(\mathcal{D}_{X,f^*\lambda}, f^*i)$ on X which by definition is associated to the data $f^*\omega \in H^1(X, \mathcal{Z}^1)$. Let $\mathcal{D}_{Y,\lambda}$-mod (resp. $\mathcal{D}_{X,f^*\lambda}$-mod) be the category of $\mathcal{D}_{Y,\lambda}$-modules on Y (resp. $\mathcal{D}_{X,f^*\lambda}$-modules on X). Then one can define two functors :

- inverse image: $f^! : \mathcal{D}_{Y,\lambda}\text{-mod} \to \mathcal{D}_{X,f^*\lambda}\text{-mod}$
- direct image: $f_* : \mathcal{D}_{X,f^*\lambda}\text{-mod} \to \mathcal{D}_{Y,\lambda}\text{-mod}$

In case f is not affine, the direct image functor has good properties only between the derived categories of bounded complexes of $\mathcal{D}$-modules. At the level of derived categories, the functor $f^!$ is simply the inverse image of $\mathcal{O}$-modules shifted by the relative dimension of f : $\mathrm{rd}(f) = \dim Y - \dim X$. One can transform a left $\mathcal{D}$-module into a right one by tensorization with Ω the sheaf of top degree differential forms. Then the direct image of a right $\mathcal{D}_{X,f^*\lambda}$-module $\mathcal{M}$ in the derived category is the direct image as $\mathcal{O}$-module of $\mathcal{M} \otimes_{\mathcal{D}_{X,f^*\lambda}} f^\circ \mathcal{D}_{Y,\lambda}$. For the material of this section we refer to [Bernstein] and [Beilinson - Bernstein 1983].

I.2 $\mathcal{D}$-modules with a group action

Let G be a complex algebraic group with Lie algebra $\mathbf{g}$, and suppose G acts on a smooth variety X. Let $\mathcal{F}$ be a sheaf on X. By definition, a *weak action* of G on $\mathcal{F}$ is an action of G on $\mathcal{F}$ which extends the action of G on X. When $\mathcal{F}$ is a $\mathcal{D}_X$-module, there is however more structure involved. If α denote the action of G on X, then we have a morphism $d\alpha : \mathbf{g} \to \Gamma(X, \mathcal{D}_X)$, where $\Gamma(X, \mathcal{F})$ denotes the global sections of $\mathcal{D}_X$, *i.e.* the ring of differential operators defined on all of X. $\Gamma(X, \mathcal{D}_X)$ acts on itself by the commutator action ad, so we get a map $ad \cdot d\alpha : \mathbf{g} \to \mathrm{End}\Gamma(X, \mathcal{D}_X)$. On the other hand $\mathcal{D}_X$ has obviously a weak action of G, say β, which yields the map $d\beta : \mathbf{g} \to \mathrm{End}\,\Gamma(X, \mathcal{D}_X)$. It is natural to require that the map $d\alpha$ be G-equivariant and that $ad \cdot d\alpha$ coincide with $d\beta$. More generally, let $(\mathcal{D}, i)$ be a sheaf of twisted differential operators on X.

2.1 Definition: An *action* of G on $\mathcal{D}$ is a weak action β together with a morphism $\pi : \mathbf{g} \to \Gamma(X, \mathcal{D})$ such that:

- π is G-equivariant with respect to Ad on $\mathbf{g}$ and β on $\Gamma(X, \mathcal{D})$,
- for $\xi \in \mathbf{g} : \operatorname{ad} \pi(\xi) = d\beta(\xi)$.

Suppose now that $\mathcal{D}$ is a tdo with a G-action given by β and π.

2.2 Definition: A $(\mathcal{D}, \mathrm{G})$-module $\mathcal{F}$ on X is a quasi-coherent $\mathcal{O}_X$-module with a structure γ of $\mathcal{D}$-module and a weak action δ of G such that:

- γ is G-equivariant with respect to β on $\mathcal{D}$ and δ on $\mathcal{F}$,
- for $\xi \in \mathbf{g} : \gamma\pi(\xi) = d\delta(\xi)$

The difference between a weak action and an action can also be illustrated as follows. Consider the diagram

$$\begin{array}{ccccc} X & \xleftarrow{p} & G \times X & \xrightarrow{a} & X \\ x & \longleftarrow & (g, x) & \longrightarrow & \alpha(g)x \end{array}$$

where a is the action morphism and p is the projection on the second factor. Then $\mathcal{D}$ has a G-action if and only if $p^\circ\mathcal{D} \cong a^\circ\mathcal{D}$ and a G-action on the $\mathcal{D}$-module $\mathcal{F}$ is equivalent to the data of an isomorphism $p^\circ\mathcal{F} \cong a^\circ\mathcal{F}$.

Decoding the definitions, one gets:

2.3 Proposition: *Suppose that G acts transitively on X and let H be the stabilizer of a point $x \in X$.*

1. *The tdo on X with a G-action are in bijection with the tdo on $\{x\}$ with an H-action, that is the 1-dimensional representations π of $\mathbf{h} = \mathrm{Lie}\, H$.*
2. *The $(\mathcal{D}, G)$-modules on X are in bijection with the $(\mathbf{C}, H)$-modules on $\{x\}$, that is the representations (V, β) of H such that for $\xi \in \mathbf{h} : \pi(\xi) = d\beta(\xi)$.*

2.4 Remark: From this proposition, we see that if H is connected, the sheaf $\mathcal{V}$ of sections of a G-homogenous vector bundle on X will not be a $(\mathcal{D}, G)$-module when $\operatorname{rank} \mathcal{V} > 1$. In fact in this case, $\mathcal{V}$ is a module over a sheaf of matrix differential operators, as in Remark 1.5. We could enlarge the theory of $\mathcal{D}$-modules to include the case of matrix differential operators acting on matrix valued functions. We will use these sheaves in chapter IV.4.

I.3 Parabolic subgroups and flag spaces

The structure theory of linear algebraic groups is clearly explained in Springer's book, so here we will only review certain facts from a point of view suited to the later developments.

The *unipotent radical* R_uG of a linear algebraic group G is the largest closed, connected, unipotent normal subgroup of G. G is called *reductive* if $R_uG = \{e\}$. Henceforth, G shall denote a complex, connected, reductive, linear algebraic group. The connectedness assumption is not at all necessary for the study of $\mathcal{D}$-modules but it simplifies the exposition here.

A *parabolic* subgroup P of G is a closed subgroup such that the quotient variety G/P is complete. A *Borel* subgroup B is a connected and solvable subgroup of G which is maximal for these properties. One proves that a closed subgroup of G is parabolic if and only if it contains a Borel subgroup. Since G is connected, a parabolic subgroup is its own normalizer. We prefer to view the flag space $X := G/P$ as the as the variety of all parabolic subgroups of G conjugate to P.

Let P be a parabolic subgroup of G and $N = N_P$ its unipotent radical. A *Levi* subgroup L of P is a closed subgroup such that the product map $L \times N \to P$ is an isomorphism of varieties. Then L is reductive, normalizes N and is the centralizer in G of a connected torus in P. We shall also say that L is a Levi subgroup of G.

If T is a maximal torus in P, there is a unique Levi subgroup of P containing T. All Levi subgroups of P are conjugate. The projection $P \to P/N$ gives an isomorphism of any Levi subgroup of P with P/N. Therefore we call P/N the *Levi factor* of P and we denote it by L_P; it is canonically attached to P. The Levi factors L_P and $L_{P'}$ of two conjugate parabolic subgroups P and P' are not necessarily canonically conjugate because we can twist a given isomorphism by an inner automorphism of L_P or $L_{P'}$.

Let $P_1 := (P, P)$ denote the commutator subgroup of P and put $T_P = P/P_1$. T_P is a maximal torus, and up to covering, it is isomorphic to the center of L_P; we call it the *Cartan factor* of P. For two conjugate parabolic subgroups P and P', the Cartan factors T_P and $T_{P'}$ are canonically conjugate. When P is a Borel subgroup, its Cartan factors coincides with its Levi factor and is isomorphic to a Cartan subgroup of G. This special case is good to bear in mind for understanding the sequel easily.

Let us denote by boldface letters the Lie algebras of the groups considered.

To define the set of roots of $\mathbf{t}_p$ in $\mathbf{g}$, we use the following trick. Choose a Levi subgroup L of P and let C be the connected component of its center. Then C acts by the adjoint action on $\mathbf{g}$ and we obtain the roots of $\mathbf{c}$ in $\mathbf{g}$. Note that these roots do not always form a root system. Now C is a finite cover of T_P by the map $C \hookrightarrow P \to T_P$, hence to every root of $\mathbf{c}$ in $\mathbf{g}$ corresponds a unique linear form on $\mathbf{t}_p$. We call these linear forms: the roots of $\mathbf{t}_p$ in $\mathbf{g}$; they are independent of the choice of L.

Let $R(\mathbf{t}_p) \subset \mathbf{t}_p^*$ be the set of roots of $\mathbf{t}_p$ in $\mathbf{g}$. $R(\mathbf{t}_p)$ is naturally divided into the set of roots whose root spaces are contained in $\mathbf{n}$ and its complement. Let $R^+(\mathbf{t}_p)$ be the set of roots of $\mathbf{t}_p$ in $\mathbf{g}/\mathbf{p}$. If α is a root of $\mathbf{t}_p$ in $\mathbf{g}$, the corresponding root space $\mathbf{g}_\alpha$ need not have dimension 1; $\dim \mathbf{g}_\alpha$ is called the multiplicity of α. Let ρ_p be the half sum of the roots contained in $R^+(\mathbf{t}_p)$ counted with their multiplicities.

Let B be a Borel subgroup of G, contained in P. The map $B/B_1 \to P/P_1$ gives a canonical surjective homomorphism $T_B \to T_P$ which dualizes to an inclusion $\mathbf{t}_p^* \to \mathbf{t}_b^*$. Hence we may think of $R(\mathbf{t}_p)$ as a subset of $\mathbf{t}_b^*$. $R(\mathbf{t}_p)$ is not always a root system; for instance, let $\mathbf{g} = D_4$ and take a Levi factor whose semisimple part is of type A_2. If we had chosen a invariant bilinear form on $\mathbf{g}^*$, then $R(\mathbf{t}_p)$ could be viewed as the projection of $R(\mathbf{t}_b)$ into $\mathbf{t}_p^*$, and the multiplicity of $\alpha \in R(\mathbf{t}_p)$ would be the number of roots in $R(\mathbf{t}_b)$ which project onto α.

We can also define the roots in the Levi factor ℓ_p as follows. Choose a maximal torus T in B. There is a unique Levi subgroup L of P containing T, so we have the roots of $\mathbf{t}$ in ℓ. From the canonical isomorphism $\mathbf{t}_b^* \xrightarrow{\sim} \mathbf{t}^*$, we obtain the set $R(\mathbf{t}_b, \ell_p)$ of roots of $\mathbf{t}_p$ in ℓ_p which is independent of the choice of T. $R(\mathbf{t}_b, \ell_p) \subseteq R(\mathbf{t}_b) \subset \mathbf{t}_b^*$. Set $R^+(\mathbf{t}_b, \ell_p) = R^+(\mathbf{t}_b) \cap R(\mathbf{t}_b, \ell_p)$ and let ρ_ℓ be the half sum of the roots in $R^+(\mathbf{t}_b, \ell_p)$. We have $\rho_\ell = \rho_b - \rho_p$.

The outcome of this stylistic exercise is that we have a canonical comparison between the various root systems considered. To a Borel subgroup B of G, we can associate the triple $(\mathbf{t}_b, R^+(\mathbf{t}_p), R(\mathbf{t}_b))$ where $\mathbf{t}_b = \mathbf{b}/\mathbf{b}_1$ and $R^+(\mathbf{t}_p) \subset \mathbf{t}_b^*$. For different Borel subgroups, these triples are canonically conjugate. So we identify them with one abstract triple $(\mathbf{t}, R^+, R)$ called the *Cartan triple* of G. The reductivity of G implies that $R = R^+ \cup -R^+$. Similarly to a pair $B \subseteq P$ we can associate the quadruple $(\mathbf{t}, \mathbf{t}_p, R^+(\mathbf{t}, \ell_p), R^+(\mathbf{t}_p))$ where $\mathbf{t}_p = \mathbf{p}/\mathbf{p}_1, R^+(\mathbf{t}_p) \subset \mathbf{t}_p^* \hookrightarrow \mathbf{t}_b^* \supset R^+(\mathbf{t}, \ell_p)$. Note that there is a canonical element $\rho_\ell = \rho_b - \rho_p$ in $\mathbf{t}^*$.

Let $D(\mathbf{t})$ be the set of simple roots in $R^+(\mathbf{t})$. The set $D(\ell_p) = D(\mathbf{t}) \cap R(\mathbf{t}, \ell_p)$ of simple roots of L characterizes the conjugacy class of the parabolic subgroup P. Via this correspondence, the subsets of $D(\mathbf{t})$ are in bijection with the G-conjugacy classes of parabolic subgroups of G.

If P is a parabolic subgroup of G corresponding to the subset I of $D(\mathbf{t})$, let $X := X_I$ the space of all subgroups of G conjugate to P. X is called the flag space of G of type P or of type I; we will also use the words flag manifold or flag variety of type P. When P is a Borel subgroup of G, we call X the full flag variety of G. X is always a complete smooth complex projective algebraic variety on which G acts transitively by conjugation. If $I \subseteq J$ are two subsets of $D(\mathbf{t})$, and if P is a parabolic subgroup of G of type I, there exists a unique parabolic subgroup Q of type J such that $P \subseteq Q$. This yields a smooth fibration $X_I \to X_J$ with fiber isomorphic to the flag variety of Q of type P.

I.4 Differential operators on flag spaces

Let X be the flag space of G of type P. Let $\mathcal{O} = \mathcal{O}_X$ be the structure sheaf of X and $\mathcal{T}_X$ be its tangent sheaf. We denote by $\alpha : \mathbf{g} \to \mathcal{T}_X$ the morphism of Lie algebras defined by the action of G on X. Let U be the enveloping algebra of $\mathbf{g}$. Put $U^\circ := \mathcal{O} \otimes_{\mathbf{C}} U$ and endow this sheaf with a multiplication extending the ring structure of U, and the structure of $\mathcal{O}$-module on U°. Explicitly for $f, g \in \mathcal{O}, A, B \in U$:

$$[f \otimes A, g \otimes B] = fg \otimes [A, B] + f\alpha(A)g \otimes B - g\alpha(B)f \otimes A$$

A direct computation shows that this bracket induces a Lie algebra structure on $\mathbf{g}^\circ := \mathcal{O} \otimes_{\mathbf{C}} \mathbf{g} \subset U^\circ$. Set:

$$\mathbf{p}^\circ = \ker(\alpha : \mathbf{g}^\circ \to \mathcal{T}_X) = \{\xi \in \mathbf{g}^\circ | \xi_x \in \mathbf{p}_x \ \forall x \in X\}.$$

Here $\mathbf{p}_x$ is the parabolic subalgebra corresponding to the point x in X and ξ_x is the value of the local section ξ at the point x. Since $\mathbf{p}^\circ$ is the kernel of a morphism of Lie algebras, it is an ideal in $\mathbf{g}^\circ$, and hence $\mathbf{p}_1^\circ$ is also an ideal in $\mathbf{g}^\circ$. Moreover the restriction of the bracket to $\mathbf{p}^\circ$ is $\mathcal{O}$-linear and $\mathbf{p}^\circ/\mathbf{p}_1^\circ \simeq \mathcal{O} \otimes_{\mathbf{C}} \mathbf{t}_p$.

We can use proposition 2.3 to classify the sheaves of twisted differential operators on X which have a G-action. They correspond precisely to the linear forms $\lambda : \mathbf{p} \to \mathbf{C}$ trivial on $\mathbf{p}_1$, i.e. to the elements of $\mathbf{t}_p^*$. As is customary, the center of symmetry of the picture is not o but $\rho_p \in \mathbf{t}_p^*$. So to avoid further normalizations, we set $\mathcal{D}_\lambda$ to be the tdo on X corresponding to the weight $\lambda - \rho_p \in \mathbf{t}_p^*$. We can describe more explicitly the tdo $\mathcal{D}_\lambda$. Every weight $\lambda \in \mathbf{t}_p^*$ determines a morphisme $\lambda^\circ : \mathbf{p}^\circ \to \mathcal{O}_X$. Let $\mathcal{I}_\lambda$ be the ideal of U° generated by the elements $\xi - (\lambda - \rho_p)^\circ(\xi)$ where ξ is a local section of $\mathbf{p}^\circ$. Then $\mathcal{D}_\lambda = U^\circ / \mathcal{I}_\lambda$.

4.1 Example: Let $\lambda \in \mathbf{t}_p^*$, i.e. $\lambda \in \mathrm{Hom}(T_P, \mathbf{C}^*)$, and let $\mathcal{O}(\lambda)$ be the corresponding invertible G-sheaf of $\mathcal{O}$-modules. Then $\mathrm{Diff}\,\mathcal{O}(\lambda) = \mathcal{D}_{\lambda+\rho_p}$. In particular $\mathcal{D}_X = \mathcal{D}_{\rho_p}$.

Let us denote by D_λ the global sections on X of $\mathcal{D}_\lambda$. The center of D_λ is $\mathbf{C}$: the constant functions. We have a morphisme $\pi : \mathbf{g} \to D_\lambda$, since $\mathcal{D}_\lambda$ is a tdo. It extends to a morphism $\pi : U \to D_\lambda$ which must send the center Z of U to $\mathbf{C} \in D_\lambda$ by a certain character θ. On the other hand the Harish-Chandra isomorphism ψ identifies Z with the ring $Z(\mathbf{t})^W$ of polynomials on $\mathbf{t}$ invariant by the Weyl group W of G. One should think of ψ first as a collection of isomorphisms, one for every point of the full flag space, which turns out to be constant once we identify the Cartan factors of all Borel subgroups of G with $\mathbf{t}$. By transposition, we obtain $\psi^* : \mathbf{t}^* \to \mathrm{Spec}\, Z : \lambda \mapsto \chi_\lambda$. It follows that $\chi_\lambda = \chi_\mu$ if and only if $\lambda = w\mu$ for some $w \in W$.

4.2 Lemma: *Let X be the flag space of type P with Levi factor L, and let $\lambda \in \mathbf{t}_p^*$. Then the character $\theta : Z \to \mathbf{C} \subseteq \mathcal{D}_\lambda$ coincides with $\chi_{\lambda+\rho_\ell}$.*

The proof of this lemma is a standard gymnastic exercise with ρ-shift. The shift by ρ_ℓ reflects the difference in the action of T on the volume forms of the full flag variety and of X.

I.5 Global sections of $\mathcal{D}_\lambda$

Our goal in this section is to describe $\Gamma(X, \mathcal{D}_\lambda)$ in terms of the enveloping algebra U. In the case of the full flag variety, this question has a nice answer, but for partial flag varieties the general situation is not clear yet. We shall first construct some bigger sheaves of algebras $\mathcal{D}_\ell$ and $\mathcal{D}_{\mathbf{t}_p}$ which will give us some insight in the problem.

Recall that $\mathbf{p}_1^\circ = \{\xi \in \mathbf{g}^\circ \,|\, \xi_x \in \mathbf{p}_{x,1} \ \forall x \in X\}$, and similarly define $\mathbf{n}^\circ = \{\xi \in \mathbf{g}^\circ \,|\, \xi_x \in \mathbf{n}_x, \forall x \in X\}$. These are both G-invariant subsheaves of

$\mathbf{g}^\circ$, hence these are ideals in $\mathbf{g}^\circ$.

5.1 Definition:

$$\mathcal{D}_\ell = U^\circ/U^\circ\mathbf{n}^\circ \qquad \mathcal{D}_{\mathbf{t}_p} = U^\circ/U^\circ\mathbf{p}_1^\circ$$

The center of $\mathcal{D}_\ell$ is isomorphic to the center $Z(\ell)$ of $U(\ell)$, while the center of $\mathcal{D}_{\mathbf{t}_p}$ is isomorphic to $Z(\mathbf{t}_p) = U(\mathbf{t}_p)$. We can view these algebras as living respectively on $\operatorname{Spec} Z(\ell) = \mathbf{t}^*/W_L$ where W_L is the Weyl group of L and $\operatorname{Spec} Z(\mathbf{t}_p) = \mathbf{t}_p^*$. Of course $\mathbf{t}_p^* = (\mathbf{t}^*)^{W_L}$ imbeds naturally into $\mathbf{t}$.$\mathcal{D}_{\mathbf{t}_p}$ is the specialization of $\mathcal{D}_\ell$ along the subvariety $\mathbf{t}_p^*$, and for $\lambda \in \mathbf{t}_p^*, \mathcal{D}_\lambda$ is the specialization of $\mathcal{D}_{\mathbf{t}_p}$ at the point λ.

Geometrically we can interpret the sheaves $\mathcal{D}_\ell$ and $\mathcal{D}_{\mathbf{t}_p}$ as follows. Identify X with G/P and let $P = LN$ be a Levi decomposition of P, recall that P_1 is the commutator subgroup of P. We have two fibrations $\pi = G/N \to G/P$ and $\tau = G/P_1 \to G/P$. Then $\mathcal{D}_\ell = \pi_\circ(\mathcal{D}_{G/N})^L$ is the direct image in the category of $\mathcal{O}$-modules of the sheaf of differential operators on G/N which commute with the right action of L on the fibers of π. Similarly, $\mathcal{D}_{\mathbf{t}_p} = \tau_\circ(\mathcal{D}_{G/P_1})^C$ where $C \simeq T_P$ is the connected component of the center of L.

We have the Harish-Chandra isomorphism for L : $\psi_L : Z(\ell) \to Z(\mathbf{t})^{W_L}$ which involves only a shift by ρ_ℓ. We also have a homomorphism $\phi : Z \to Z(\ell)$ which involves a shift by ρ_p. Let us denote by H the space of harmonic polynomials on $\mathbf{t}$ for the action of the Weyl group W of $\mathbf{g}$; $\dim H(\mathbf{g}) = |W|$: the cardinality of W. A similar notation holds for L and $H(\ell) \subseteq H$. Chevalley's theorem asserts that:

$$Z(\mathbf{t}) = H \otimes Z \qquad Z(\mathbf{t}) = H(\ell) \otimes Z(\ell)$$

It is readily seen that $Z(\ell) = H^{W_L} \otimes Z$, so that $Z(\ell)$ is a free Z-algebra on $|W/W_L|$ generators. Indeed H is isomorphic to the regular representation of W and H^{W_L} is isomorphic to the space of functions on W/W_L.

5.2 Definition: $U_\ell = U \otimes_Z Z(\ell)$

There is a natural map $Z(\ell) \to \mathcal{D}_\ell$, because although $\mathbf{p}^\circ/\mathbf{n}^\circ$ is not necessarily a trivial bundle over X, the ambiguity disappears when we consider the center of the enveloping algebra of ℓ. Moreover the restriction of this map to Z coincides with the restriction of $U \to \mathcal{D}_\ell$ to Z. Hence we obtain a well-defined morphisme $U_\ell \to \Gamma(X, \mathcal{D}_\lambda)$.

5.3 Lemma:

$$U_\ell \longrightarrow \Gamma(X, \mathcal{D}_\ell) \text{ is an isomorphism}$$
$$H^i(X, \mathcal{D}_\ell) = 0 \text{ for } i > 0.$$

The proof we will sketch generalizes [Miličič] proof for the full flag variety. First observe that $Z(\ell) \to \mathcal{D}_\ell$ is an isomorphism onto the center of $\mathcal{D}_\ell$. Next to show that $U \otimes_Z Z(\ell) \xrightarrow{\sim} \Gamma(X, \mathcal{D}_\ell)$, it suffices to prove that it is an isomorphism at the graded level. Set $S = S(\mathbf{g})$, then $S^G = Z$. Put $Y = G/N, X = G/P, \pi : T^*Y \to X$ and let $\mathcal{O}^L_{T^*Y}$ be the sheaf of regular functions on T^*Y which are right invariant under the action of L on Y and which are homogenous in the fiber variables of the projection $T^*Y \to Y$. Then $\mathrm{gr}\mathcal{D}_\ell \simeq \pi_\circ \mathcal{O}^L_{T^*Y}$ and $\Gamma(X, \pi_\circ \mathcal{O}^L_{T^*Y}) = R(T^*Y)^L$ is the ring of regular functions on T^*Y which are right L-invariant. Since there is a natural inclusion $\mathrm{gr}\Gamma(X, \mathcal{D}_l) \hookrightarrow \Gamma(X, gr\mathcal{D}_l)$, it suffices to prove that:

$$S \otimes_Z Z(\ell) \xrightarrow{\sim} R(T^*Y)^L$$

To prove this, one resolves the sheaf $\pi_\circ \mathcal{O}^L_{T^*Y}$ by a Koszul complex $\mathcal{C} = S^\circ \otimes_{\mathcal{O}} \bigwedge^\bullet \mathbf{n}^\circ$; $\mathcal{C}$ has an obvious structure of left $\mathbf{g}$-module, but it has also a structure r of right $\mathbf{g}$-module via the formula:

$$r(x)(u \otimes v) = -ux \otimes v + u \otimes [x, v]$$

for $x \in \mathbf{g}, u \in S^\circ$ and $v \in \bigwedge^\bullet \mathbf{n}^\circ$. $\mathcal{C}$ is endowed with the usual derivation which preserves the $(\mathbf{g}, \mathbf{g})$-module structure. One proves that this $\mathcal{C}$ complex is acyclic. Moreover, $S^\circ / S^\circ \mathbf{n}^\circ$ is equal to $\mathrm{gr}\mathcal{D}_\ell$, thus $\mathcal{C}$ is a left resolution of the $(\mathbf{g}, \mathbf{g})$-module $\mathcal{D}_\ell$. There is a third quadrant spectral sequence whose term $E_1^{p,-q}$ is $H^p(X, S^\circ \otimes_{\mathcal{O}} \bigwedge^q \mathbf{n}^\circ)$ and which abuts to its term $E_\infty^{p,-q} = H^{p-q}(X, \mathrm{gr}\mathcal{D}_\ell)$.

Now we can compute $H^p(X, S^\circ \otimes_{\mathcal{O}} \bigwedge^q \mathbf{n}^\circ) = S \otimes H^p(X, \bigwedge^q \mathbf{n}^\circ)$. Indeed we can identify $\bigwedge^q \mathbf{n}^\circ$ with Ω^q_X the sheaf of holomorphic q-forms on X via the Killing form on $\mathbf{g}$. By Dolbeault's theorem $H^p(X, \Omega^q) \simeq H^{q,p}_{\bar\partial}(X)$, and since X is a compact Kähler smooth manifold, the Hodge theorem implies that $H^n(X, \mathrm{C}) \simeq \bigoplus_{p+q=n} H^{q,p}_{\bar\partial}(X)$. On the other hand, the cohomology of a flag variety is generated by the fundamental classes of the Schubert cycles. Algebraic cycles live only in degrees $p = q$. There is a natural length function on the quotient W/W_L; let us denote by ℓ_p the number of elements of length p in W/W_L. Then we obtain:

$$H^p(X, \Lambda^q \mathbf{n}^\circ) = 0 \ , \ \text{if } p \neq q$$
$$\dim H^p(X, \Lambda^p \mathbf{n}^\circ) = \ell_p$$

This implies that the spectral sequence degenerates, and that:

$$E^n_\infty = H^n(X, \operatorname{gr} \mathcal{D}_\ell) = 0 \text{ for } n \neq 0$$
$$\operatorname{gr} E^0_\infty = \operatorname{gr}\Gamma(X, \operatorname{gr} \mathcal{D}_\ell) = S \otimes H^*(X, \mathbf{C})$$

$H^*(X, \mathbf{C})$ is nothing else than the space of W_L-invariant harmonic polynomials on $\mathbf{t}^*$ for the full Weyl group W. It follows, using Chevalley's theorem recalled above, that $S \otimes_Z \operatorname{gr} Z(\boldsymbol{\ell})$ is isomorphic to $\operatorname{gr}\Gamma(X, \operatorname{gr} \mathcal{D}_\ell)$. And since the gradations correspond, we obtain the desired result. □

I have not found a simple description of $\Gamma(X, \mathcal{D}_{\mathbf{t}_p})$, but at least the following can be said. Let $X_1 = G/P_1$, and let $D(X_1)$ denote the algebra of algebraic invariant differential operators on X_1. The action of G on X_1 gives a morphism $op : U \to D(X_1)$ called the operator representation of U on X_1. Let J_P denote its kernel.

5.4 Proposition: (Borho-Brylinski)

$$J_P = \operatorname{Ann}(U \otimes_{U([\mathbf{p},\mathbf{p}])} \mathbf{C}) = \bigcap_{\lambda \in \mathbf{t}_p^*} \operatorname{Ann}(U \otimes_{U(\mathbf{p})} \mathbf{C}_\lambda) .$$

The map $\pi : X_1 \to X$ is a G-equivariant T_P-fibration. The ring $\pi_\circ \mathcal{O}_{X_1}$ is graded by the lattice of characters of T_P acting on the right of X_1 and we have the corresponding gradation on $\pi_\circ \mathcal{D}_{X_1}$. The zero component is just $\mathcal{D}_{\mathbf{t}_p}$, and since the G-action commutes with the right T_P-action, $\pi_\circ op(U)$ lies in $\Gamma(X, \mathcal{D}_{\mathbf{t}_p})$. On the other hand the right action of T_P on every fiber of π gives a monomorphism $r : Z(\mathbf{t}_p) \to D(X_1)$, and by the commutativity of T_P, $\pi_\circ r Z(\mathbf{t}_p)$ lies also in $\Gamma(X, \mathcal{D}_{\mathbf{t}_p})$.

Now composing the Harish-Chandra homomorphism $\psi : Z \to Z(\mathbf{t})$ with the natural projection $Z(\mathbf{t}) \to Z(\mathbf{t}_p)$, we can view $Z(\mathbf{t}_p)$ as a Z-module. Then it is not difficult to see that there is a well-defined monomorphism:

$$U/J_P \bigotimes_Z Z(\mathbf{t}_p)$$

Note that when X is the full flag variety, $I(X_1) = 0$, and the above map is an isomorphism by Proposition 5.2.

Now we examine the global sections of $\mathcal{D}_\lambda, \lambda \in \mathbf{t}_p^*$. Recall that λ determines a character $\chi_\lambda : Z \to \mathbf{C}$. Let I_λ be the 2-sided ideal in U generated by the kernel of X_λ. Using Lemma 5.3, it is easy to prove:

5.5 Proposition: (Beilinson-Bernstein-Brylinski-Kashiwara) *If X is the full flag variety, for any $\lambda \in \mathbf{t}_p^*$, we have:*

$$U/I_\lambda \xrightarrow{\sim} \Gamma(X, \mathcal{D}_\lambda)$$

5.6 Proposition: *Let X be a flag space of type P and $\lambda \in \mathbf{t}_p^*$. If $\lambda + \rho_\ell$ is dominant in $\mathbf{t}^*$, then $U \to \Gamma(X, \mathcal{D}_\lambda)$ is surjective with kernel $I_\lambda + J_P$.*

Proof: Consider the map $\pi : Y \to X$ where Y is the full flag variety. $\pi^! \mathcal{D}_\lambda$ is a $\mathcal{D}_{Y,\lambda+\rho_\ell}$-module on X. By applying $\mathcal{D}_{Y,\lambda+\rho_\ell}$ to the generating section $\pi^!(1)$, we get a surjective map of sheaves $\mathcal{D}_{Y,\lambda+\rho_l} \to \pi^\circ \mathcal{D}_\lambda$. Now the functor $\Gamma(Y, \cdot)$ of global sections is exact because $\lambda + \rho_\ell$ is dominant, *cf.* theorem 6.3. Hence $\Gamma(Y, \mathcal{D}_{Y,\lambda+\rho_l}) \to \Gamma(Y, \pi^\circ \mathcal{D}_\lambda)$ is still surjective. But U surjects onto the first algebra by the previous result, and $\Gamma(Y, \pi^\circ \mathcal{D}_\lambda) = \Gamma(X, \mathcal{D}_\lambda)$. Thus U surjects onto $\Gamma(X, \mathcal{D}_\lambda)$. The assertion on the kernel is clear from the discussion above.

5.7 Remark: $U/(I_\lambda + J_P)$ always injects into $\Gamma(X, \mathcal{D}_\lambda)$. But if $\lambda + \rho_\ell$ is not dominant in $\mathbf{t}^*$, then $U \to \Gamma(X, \mathcal{D}_\lambda)$ may not be surjective, even though λ is dominant in $\mathbf{t}_p^*$. The simplest example (due to W. Borho) of this phenomenon occurs for a partial flag variety of the group SP_2 (4×4 matrices). In this case, the moment map is not birational (see III.5 and IV.1 for definitions). One detects the non-surjectivity by looking at an irreducible $\Gamma(X, \mathcal{D}_\lambda)$-module whose restriction to $U(sp_2)$ is a reducible highest weight module.

I.6 $\mathcal{D}$-affine varieties

The theory of Beilinson and Bernstein works as well over any algebraically closed field k of characteristic zero. So let X be a scheme over k. Define an $\mathcal{O}_X$-ring $\mathcal{R}$ to be a sheaf of rings on X together with a ring morphism $\mathcal{O}_X \to \mathcal{R}$ such that $\mathcal{R}$ is quasicoherent as a left $\mathcal{O}_X$-module. An $\mathcal{R}$-module is then a sheaf of left $\mathcal{R}$-modules, quasicoherent as a sheaf of $\mathcal{O}_X$-modules. Denote by $\mathcal{R}$-mod the category of $\mathcal{R}$-modules. Put $R := \Gamma(X, \mathcal{R})$. There are natural adjoint functors $\Gamma : \mathcal{R}-\text{mod} \leftrightarrows R-\text{mod} : \Delta$, where $\Gamma(\mathcal{M}) := \Gamma(X, \mathcal{M})$ are the global sections of $\mathcal{M}$ and $\Delta(N) := \mathcal{R} \otimes_R N$ is called the *localization* of N. Γ is left exact, Δ is right exact and we have the derived functors $R\Gamma$ and $L\Delta$.

6.1 Definition: We say that X is $\mathcal{R}$-affine if Γ and Δ are (mutually inverse) equivalence of categories.

Here is a criterion for $\mathcal{R}$-affinity.

6.2 Proposition: *If every $\mathcal{R}$-module is generated by its global sections and $H^i(X, \mathcal{M}) = 0$ for $i > 0$, then X is $\mathcal{R}$-affine.*

This proposition says that if Γ is exact and faithful, then it is an equivalence of categories. This is a consequence of the Gabriel–Mitchell theorem in theory of categories, see [Bass, p. 54]. It is clear by Serre's theorem that any affine variety is $\mathcal{R}$-affine. $\mathcal{D}_X$ is an $\mathcal{O}_X$-ring; we shall see that any flag space X is $\mathcal{D}_X$-affine. $\lambda \in \mathbf{t}_p$, of twisted differential operators on a flag space X of type P for G. For simplicity we consider only the case $k = \mathbf{C}$. The $d\lambda$ for $\lambda \in \text{Mor}(T_P, \mathbf{C}^\times)$ define a lattice in $\mathbf{t}_p^*$ and hence a real structure. We shall say that $\lambda \in \mathbf{t}_p^*$ is *P-dominant* if for any root $\alpha \in R^+(\mathbf{t}_p)$ we have $< \lambda, \alpha^\vee > \neq -1, -2, \ldots$. We shall say that $\lambda \in \mathbf{t}_p^*$ is *P-regular* if for any root $\alpha \in R^+(\mathbf{t}_p)$, we have $< \lambda, \alpha^\vee > \neq 0$. Recall that the positive roots are those which are in $\mathbf{g}_p$. Via the inclusion $\mathbf{t}_p^* \hookrightarrow \mathbf{t}^*$, we view the elements of $\mathbf{t}_p^*$ as elements of $\mathbf{t}^*$, and there is a well-defined element $\rho_\ell = \rho_b - \rho_p \in \mathbf{t}^*$. It is easy to check that if $\lambda \in \mathbf{t}_p^*$ is B-dominant, then it is also P-dominant, and vice-versa. But if $\lambda \in \mathbf{t}_p^*$ is P-dominant, it may not be true that $\lambda + \rho_\ell \in \mathbf{t}^*$ is B-dominant, see exercise 6.7 below. Put $D_\lambda = \Gamma(X, \mathcal{D}_\lambda)$.

6.3 Theorem: (Beilinson-Bernstein)

1. *If $\lambda \in \mathbf{t}_p^*$ is P-dominant, then the functor $\Gamma : \mathcal{D}_\lambda\text{-mod} \to D_\lambda\text{-mod}$ is exact.*

2. *If $\lambda \in \mathbf{t}_p^*$ is P-dominant and $\lambda + \rho_\ell \in \mathbf{t}^*$ is B-regular, then the functor Γ is also faithful.*

Thus, under the conditions of the theorem, X is $\mathcal{D}_\lambda$-affine. The case of the full flag variety is explained in [Beilinson-Bernstein 1981] and this theorem can be proved in an similar way. We will only describe the changes for the key lemma, and show a consequence of the sole exactness of Γ.

Set $\mathcal{O} = \mathcal{O}_X$. Let F be an irreducible G-module, $\mathcal{F} = \mathcal{O} \otimes_{\mathbf{C}} F$ the corresponding G-sheaf. Let $(\mathcal{F}_i)$, for $i = 1, \ldots, k$, be a filtration of $\mathcal{F}$ by G-sheaves of $\mathcal{O}$-modules such that the quotients $\mathcal{F}_i/\mathcal{F}_{i-1} \simeq \mathcal{V}(\mu_i)$ correspond to irreducible representations of the Levi factor L of P with highest weight $\mu_i \in \mathbf{t}^*$. If $\mu \notin \mathbf{t}^*, \mathcal{V}(\mu)$ is not a $\mathcal{D}$-modulestrictly speaking because it is not an invertible sheaf. It is the sheaf of sections of a homogenous vector bundle over X and what really acts on $\mathcal{V}(\mu)$ are twisted matrix differential operators. The ring of these operators is still generated locally by $\mathcal{O}$ and the enveloping algebra U. Hence $\nu(\mu)$ is a U°-module and the center Z of U acts on it by the character $\chi_{\mu+\rho_b}$. Let us call *L-weights* of F the weights $\mu \in \mathbf{t}^*$ which appear in the Jordan-Hölder series of $\mathcal{F}$. Let μ be the highest weight of $\mathcal{F}$ and ν be the highest weight of F^*. Set $\mathcal{F}(\mu) = \mathcal{V}(\mu) \otimes_{\mathcal{O}} \mathcal{F}$. Let $i : \mathcal{F}_1(\nu) \to \mathcal{F}(\nu)$ and $p : \mathcal{F} \to \mathcal{F}_k/\mathcal{F}_{k-1} \simeq \mathcal{V}(\mu)$. For any $\mathcal{O}$-module $\mathcal{M}$, put $i_{\mathcal{M}} = i \otimes id_{\mathcal{M}} : \mathcal{F}_1(\nu) \otimes \mathcal{M} \to \mathcal{F} \otimes \mathcal{M}(\nu)$ and $p_{\mathcal{M}} = p \otimes id_{\mathcal{M}} : \mathcal{F} \otimes \mathcal{M} \to \mathcal{M}(\mu)$. Observe that if $\nu \in \mathbf{t}_p^*$, then $\mathcal{F}_1(\nu) \simeq \mathcal{O}$, and $i_{\mathcal{M}} : \mathcal{M} \to \mathcal{F} \otimes \mathcal{M}(\nu)$. Now let $\mathcal{M}$ be a $\mathcal{D}_\lambda$-module, then all the sheaves considered above have a structure of U°-modules by the Leibnitz formula and $i_{\mathcal{M}}$, $p_{\mathcal{M}}$ are morphisms of U°-modules.

6.4 Lemma:

1. *Take $\nu \in \mathbf{t}_p^*$. If $\lambda \in \mathbf{t}_p^*$ is P-dominant, then $i_{\mathcal{M}}$ has a right inverse $j_{\mathcal{M}}$ (unique) in the category of U°-modules.*

2. *Take $\mu \in \mathbf{t}_p^*$. If in addition $\lambda + \rho_\ell \in \mathbf{t}^*$ is B-regular, then $p_{\mathcal{M}}$ has a right inverse $q_{\mathcal{M}}$ (unique) in the category of U°-modules.*

Proof: Consider the filtration $\mathcal{F}_i \otimes \mathcal{M}(\nu)$ of $\mathcal{F} \otimes \mathcal{M}(\nu)$. It is easy to check that the subquotients $\mathcal{F}_i \otimes \mathcal{M}(\nu)/\mathcal{F}_{i-1} \otimes \mathcal{M}(\nu) = \mathcal{M}(\mu_i + \nu)$ are U°-modules on which Z acts by the characters $\chi_i = \chi_{\lambda+\mu_i+\nu+\rho_\ell}$.

Claim: The weight $\lambda + \rho_\ell = \lambda + \mu_1 + \nu + \rho_\ell \in \mathbf{t}^*$ is not conjugate by the Weyl group of G to any weight $\lambda + \mu_i + \nu + \rho_\ell$ for $i > 1$.

To prove such a statement, we can assume that λ is dominant in the analytic sense,i.e. $< Re\lambda, \alpha^\vee > \geq 0$ for all $\alpha \in R^+(\mathbf{t}_p)$. Indeed, the principle explained in the appendix of [Bernstein-Gelfand] will take care of the passage from the analytic notion of dominance to the algebraic one. Now suppose that there exists some element $w \in W$ such that

$$w(\lambda + \rho_\ell) = \lambda + \mu_i + \nu + \rho_\ell, \text{ for some } i,$$

Since $\mu_1 \in \mathbf{t}_p^*$ is the lowest weight of F with respect to $\mathbf{t}, -\mu_1 + \mu i$ is a sum of positive roots. Let us introduce a norm $|\cdot|$ on $\mathbf{t}^*$ coming from a W-invariant scalar product and let $|\cdot|_N$, resp. $|\cdot|_L$ denote the composition of the projections onto $\mathbf{t}_p^*$, resp. $\mathbf{t}_p^* \perp$, followed by the norm. We have $|x| = |x|_N + |x|_L$ for $x \in \mathbf{t}$. Then $|\lambda| = |\lambda|_N$ and $|\rho_\ell| = |\rho_\ell|_L$. Since W acts by isometries, if $\lambda + \rho_\ell$ is conjugate to $\lambda + \mu_i + \nu + \rho_\ell$, they must have the same norm. But

$$|\lambda + \rho_\ell|_N < |\lambda + \mu_i + \nu + \rho_\ell|_N \text{ unless } \mu_i = \mu_1,$$
$$|\lambda + \rho_\ell|_L \leq |\lambda + \mu_i + \nu + \rho_\ell|_L.$$

The first inequality follows from the fact that $\nu = -\mu_1 \in \mathbf{t}_p^*$ is the lowest weight of F and λ is P-dominant, hence B-dominant. The second inequality comes from the fact that μ_i is the highest weight of a representation of L. This proves the claim.

By Harish-Chandra's theorem, this means that $\mathcal{M} = \mathcal{F}_1 \otimes \mathcal{M}(\nu)$ is exactly the χ_1-eigenspace of Z in $\mathcal{F} \otimes \mathcal{M}(\nu)$. Therefore there exists a unique projection $j_\mathcal{M} : \mathcal{F} \otimes \mathcal{M}(\nu) \to \mathcal{M}$ which is the right inverse of $i_\mathcal{M}$ in the category of U°-modules.

The proof of the second part of the lemma is the same as for the full flag variety and we will skip it.

To finish the proof of the theorem, one uses Theorems A and B of Cartan-Serre, as it is done for the full flag variety. □

6.5 Remark: The localization functor $\Delta_\lambda : D_\lambda\text{-mod} \to \mathcal{D}_\lambda\text{-mod} : \mathcal{M} \mapsto \mathcal{D}_\lambda \otimes_{D_\lambda} M$ is left adjoint to Γ. A representation M of $\mathbf{g}$ which arises from a $\mathcal{D}$-module on X must satisfy certain conditions if X is not the full flag variety. First its infinitesimal character must be of the form $\lambda + \rho_\ell$ for some $\lambda \in \mathbf{t}_p^*$, and also its associated variety, *cf.* Chapter 3, must be contained in the closure of the Richardson orbit of P. Let $x \in X$. We can identify canonically the center of ℓ_x with $\mathbf{t}_x$. Then, $\mathbf{t}_x$ acts on the homology spaces $H_\bullet(\mathbf{n}_x, M)$, and the spectrum of its action is contained in the intersection of the W-orbit of $\lambda - \rho_p$ with $\mathbf{t}_x^*$. Denote by a subscript $\lambda - \rho_p$ the corresponding eigenspace. Then in the situation of the theorem, the geometric fiber of the sheaf $\Delta_\lambda(M)$ at x is

$$\mathbf{C}_x \otimes_{\mathcal{O}_x} \Delta_\lambda(M) = H_0(\mathbf{n}_x, M)_{\lambda-\rho_\ell}$$

and

$$\mathrm{Tor}_i^{\mathcal{O}_x}(C_x, \Delta_\lambda(M)) = H_i(\mathbf{n}_x, M)_{\lambda-\rho_\ell}\,.$$

We will often work in this book with the case where Γ is exact but not faithful. Therefore the following result will be very useful.

6.6 Proposition *Suppose* $\Gamma : \mathcal{D}_\lambda$-mod $\to D_\lambda$-mod *is exact, then*

$$\Gamma \circ \Delta = Id\,.$$

Furthermore,

(1) Γ *sends simple objects to simple ones or to zero.*
(2) Γ *sends distinct simple objects to distinct ones or to zero.*

Proof: $\Gamma \circ \Delta$ is a right exact functor which commutes with direct sums. Therefore its effect on the category D_λ-mod is determined by its action on the projective objects. Indeed given an object $M \in D_\lambda$-mod, to compute $\Gamma \circ \Delta(M)$, we can replace M by a projective resolution. In the category D_λ-mod, we may even use a two-step free resolution of the form

$$\oplus^m D_\lambda \to \oplus^n D_\lambda \to M \to 0\,.$$

The definitions of the functors imply $\Gamma \circ \Delta(D_\lambda) = \Gamma(X, \mathcal{D}_\lambda \otimes_{D_\lambda} D_\lambda) = \Gamma(X, \mathcal{D}_\lambda)$
$= D_\lambda$. Thus $\Gamma \circ \Delta$ is the identity on the whole category D_λ-mod.

To prove part (1), recall the adjunction formulai

$$\mathrm{Hom}_D(N, \Gamma\mathcal{M}) = \mathrm{Hom}_{\mathcal{D}}(\Delta N, \mathcal{M})$$

where $\mathcal{M} \in \mathcal{D}_\lambda$-mod and $N \in D_\lambda$. Let $\mathcal{M}$ be an irreducible $\mathcal{D}$-module, and suppose N is a D-submodule of $\Gamma\mathcal{M}$. By adjointness, the nonzero map $N \to \Gamma\mathcal{M}$ gives a nonzero map $\Delta N \to \mathcal{M}$ which therefore must be epi because $\mathcal{M}$ is irreducible. Applying Γ we get a nonzero map $\Gamma\Delta N \to \Gamma\mathcal{M}$ which is still epi by the exactness of Γ. But there is a natural morphism

$N \to \Gamma\Delta N$, adjoint to the identity on ΔN, and the composite map $N \to \Gamma\mathcal{M}$ is our original map. Thus, it is epi. This implies the irreducibility of ΓM.

For part (2), consider an irreducible module $M \in D_\lambda$-mod. Then ΔM has a Jordan-Hölder series whose factors are – say – $\mathcal{M}_1, \mathcal{M}_2, \ldots, \mathcal{M}_k$. Since $\Gamma \circ \Delta(M) = M$ is irreducible, only one composition factor – say $\mathcal{F}_M = \mathcal{M}_i$ – has nonzero global sections. Moreover $\mathcal{F}_M$ is an irreducible $\mathcal{D}_\lambda$-module such that $\Gamma(\mathcal{F}_M) = M$. Hence we can set up a bijection between the simple objects in D_λ-mod, and those in $\mathcal{D}_\lambda$-mod which have nonzero global sections, by sending M to $\mathcal{F}_M$. It follows that if $\mathcal{F}$ and $\mathcal{G}$ are two irreducible inequivalent $\mathcal{D}_\lambda$-modules such that $\Gamma\mathcal{F}$ and $\Gamma\mathcal{G}$ are both nonzero, then these latter D_λ-modules are distinct. □

In other words, if the functor Γ is exact, it gives an equivalence of categories between the quotient of $\mathcal{D}_\lambda$-mod by the kernel of Γ, and the category D_λ-mod.

6.7 Exercise: To become familiar with the difference between between B-dominant weights and P-dominant weights, the reader should compute a few examples. An interesting one occurs for $G = SP_4$ (8×8) matrices and P the complexification of a parabolic subgroup which is minimal in the real form $SP(1,3)$. Then $L = GL_2 \times SP_2$, and T_P is a one-dimesional torus. In usual coordinates, $\mathbf{t}_p^* = \{\lambda = (x, x, 0, 0) | x \in \mathbf{C}\}$ and $\rho_l = (\frac{1}{2}, \frac{-1}{2}, 2, 1)$. Then λ is P-dominant if $x \geq 0$, but $\lambda + \rho_l \in \mathbf{t}^*$ is B-dominant only if $x \geq \frac{5}{2}$.

I.7 Holonomic modules and Harish-Chandra modules

Let $\mathcal{D}$ be a tdo on X. A $\mathcal{D}$-module is called *smooth* if it is coherent as a sheaf of $\mathcal{O}$-module or equivalently if after a local isomorphism $\mathcal{D} = \mathcal{D}_X$ it becomes the sheaf of local sections of a vector bundle with integrable connection. One says that a $\mathcal{D}$-module is *coherent* if it is locally finitely generated. Let $\mathcal{D}$-mod c denote the category of coherent $\mathcal{D}$-module. Consider $\mathcal{D}$ as a $(\mathcal{D}, \mathcal{D})$-bimodule a natural functor of duality $* : \mathcal{D}\text{-mod}\, c \to \mathcal{D}\text{-mod}\, c$ by the formula $*\mathcal{M} = R\,\mathrm{Hom}(\mathcal{M}, \mathcal{D}[\dim X])$. One has $** = id$. A smooth $\mathcal{D}$-module$\mathcal{V}$ is of course coherent and $*\mathcal{V} = \mathrm{Hom}_{\mathcal{O}}(\mathcal{V}, \Omega_X)$ with an obvious $\mathcal{D}^o$-module structure. Coherent $\mathcal{D}$-modules correspond to finitely generated representations and $*$ corresponds to the functor $*M := R\,\mathrm{Hom}(M, U_{\chi}[\dim X])$. This duality is different from the usual contragredient functor for representations and the precise relation between the two is not transparent.

Next consider a locally closed affine imbedding $i : Y \hookrightarrow X$ with Y smooth. We denote by $\mathcal{D}_{(Y)}$ the tdo on Y inverse image of $\mathcal{D}$ on X. Let $\mathcal{M}$ be a smooth $\mathcal{D}_{(Y)}$-module on Y, then $i_*\mathcal{M}$ is a coherent $\mathcal{D}$-moduleon X. Put $i_! = *i_* * \mathcal{M}$. We have $i^! i_! \mathcal{M} = i^! i_* \mathcal{M} = \mathcal{M}$ and there is a unique morphism $f : i_!\mathcal{M} \rightarrow i_*\mathcal{M}$ such that $i^!(f) = id_{\mathcal{M}}$. Denote $\mathrm{Im}\, f$ by $i_{!*}\mathcal{M}$. If $\mathcal{M}$ is irreducible, then $i_{!*}\mathcal{M}$ is the unique irreducible submodule of $i_*\mathcal{M}$ and the unique irreducible quotient of $i_!\mathcal{M}$ (and the unique irreducible subquotient of any of these $\mathcal{D}$-modules whose restriction to Y is non-zero). The modules $i_*\mathcal{M}$, $i_!\mathcal{M}$ are called *standard modules* or also respectively maximal and minimal extension of $\mathcal{M}$, while $i_{!*}\mathcal{M}$ is called the *irreducible module* corresponding to $(Y, \mathcal{M})$ or also the middle extension of $\mathcal{M}$.

By definition a $\mathcal{D}$-module is *holonomic* if it has finite length and all its Jordan-Hölder components are irreducible modules of the type constructed above. One says that a holonomic $\mathcal{D}$-module(on compact X in the twisted case) has *regular singularities* (RS for short) if all its components originate in bundles with regular singularities at infinity. The basic property of holonomic modules is that the corresponding derived category of complexes with holonomic cohomology is stable under the functors of type $f^!, f_*$; if $\mathcal{M}$ is holonomic then $*\mathcal{M}$ is also holonomic. The same applies to holonomic RS.

Let us return to representations of $\mathbf{g}$. We are going to study $(\mathbf{g}, K)$-modules for certain algebraic groups K such that the connected component K^o is equipped with a map into G, which generally is an inclusion. First we say that $(\mathbf{g}, K)$ is a *Harish-Chandra pair* if $\mathbf{g}$ is a complex Lie algebra, K is a complex linear algebraic group (possibly disconnected) such that $\mathbf{k}$ is a subalgebra of $\mathbf{g}$ and there is a compatible map $\mathrm{Ad} : K \rightarrow \mathrm{Int}\mathbf{g}$. A *$(\mathbf{g}, K)$-module M* is by definition a representation of $\mathbf{g}$ and an algebraic representation of K on the same linear space M such that the representations coincide on $\mathbf{k}$ and the map $\mathbf{g} \times M \rightarrow M$ is K-equivariant. $(\mathbf{g}, K)$-modules correspond via the functor of localization to $(\mathcal{D}, K)$-modules, i.e. $\mathcal{D}$-modules $\mathcal{M}$ with an action of K such that $\mathbf{k}$ acts via the imbedding $\mathbf{k} \rightarrow \mathcal{D}$ and the map $\mathcal{D} \times \mathcal{M} \rightarrow \mathcal{M}$ is K-equivariant. The case where K does not act by inner automorphisms on $\mathbf{g}$ is also interesting. To include it in this framework, it suffices to assume that the group G' generated by G and the outer automorphisms of $\mathbf{g}$ given by K, acts on the flag space X and on the tdo $\mathcal{D}$. By allowing the map $K^\circ \to G$ to be different than an inclusion, we can study representations of non-linear real reductive groups.

To get an interesting theory one needs sufficiently large groups K. Say that K is *admissible* if K acts on the full flag variety of G with finitely many orbits or, equivalently, if $\mathbf{k}$ is transverse to some Borel subalgebra. Fix an admissible K. It is not hard to see that any coherent $(\mathcal{D}, K)$-module is smooth along the stratification given by the orbits of K, so is holonomic and has regular singularities. The irreducible $(\mathcal{D}, K)$-modules are in bijective correspondence via the $i_{!*}$ construction with the irreducible smooth $(\mathcal{D}_{(Y)}, K)$-modules on the various affinely imbedded K-orbits Y, and these smooth modules are charaterized by representations of the stabilizers of points. This readily gives a classification of irreducible $(\mathcal{D}, K)$-modules and so of $(\mathbf{g}, K)$-modules. For any K-orbit Y, put $T_Y = K \cap P_y / K \cap (P_y, P_y) \subseteq T_P$ where $y \in Y$ (note that T_y does not depend on $y \in Y$). T_Y is the product of the torus T_Y° and the finite abelian group T_Y/T_Y°.

7.1 Theorem: (Beilinson-Bernstein) *For $\lambda \in \mathbf{t}_p^*$, the irreducible $(\mathcal{D}_\lambda, K)$-modules are in bijective correspondence with the set of pairs (Y, τ_Y) where Y is a K-orbit in X and τ_Y is an irreducible $(\mathbf{t}, T_Y)$-module on which $\mathbf{t}$ acts by $\lambda - \rho_p$. If X is the full flag variety and if $\lambda \in \mathbf{t}$ is dominant regular, this is also the classification of irreducible $(\mathbf{g}, K)$-modules with infinitesimal character χ_λ.*

On a partial flag variety X of type P, under the hypothesis that $\lambda \in \mathbf{t}_p^*$ is P-dominant and $\lambda + \rho_\ell \in \mathbf{t}^*$ is B-regular, we obtain the classification of a certain subcategory – it often is the whole category even – of $(\mathbf{g}, K)$-modules with infinitesimal character $\chi_{\lambda+\rho_\ell}$ and associated variety contained in the closure of the Richardson orbit of P.

To work cleanly with the standard modules one has to suppose that the K-orbits Y are affinely imbedded. We see that the standard and irreducible modules corresponding to an orbit Y form families with $\dim(\mathbf{t}_p/\mathbf{t}_Y)$ continuous parameters and $\dim T_Y$ discrete parameters. All standard modules are irreducible for generic values of the parameters. If they are irreducible for all values of the parameters then Y is a closed orbit.

The groups K we will consider are the fixed points of involutions of G. This corresponds to the Harish-Chandra modules or representations of real reductive groups. Then the standard modules i_* correspond to the standard representations used in [Vogan]'s book. One could also take $K = N$ or B where B is a Borel subgroup of G and N is its unipotent radical. This corresponds to representations of $\mathfrak{g}$ with highest weights. These cases can be reduced to the first – although in general one prefers to go the other way – thanks to the facts that N-orbits are equal to B-orbits and B-orbits on X are in bijection with G-orbits on $X \times X$. This yields the equivalence between representations with highest weights and representations of complex reductive groups.

In general the hypothesis in theorem 6.3.2. can be weakened. For a $(\mathcal{D}_\lambda, K)$-module $\mathcal{M}$ with K is reductive, if $\lambda \in \mathfrak{t}_p^*$ is P-dominant and $\lambda + \rho_\ell \in \mathfrak{t}^*$ is regular with respect to the roots of K, then $\Gamma(X, \mathcal{M}) \neq 0$. This follows from the Borel-Weil theorem.

In the following chapters, we will only consider coherent $(\mathcal{D}, K)$-modules and finitely generated $(\mathfrak{g}, K)$-modules. So we add this hypothesis to the definitions. For any map $f : Y \to X$ between smooth algebraic varieties, we define the functor $f^* = *f^!*$. This chapter was meant to be only a summary. For further details on $\mathcal{D}$-modules, the reader may consult [Borel et al.] and a forthcoming book by D. Miličić.

7.2 Exercise: The reader is urged to compute some simple examples which will make this theory more concrete. The first situation to master is the construction and classification of irreducible representations of compact Lie groups. Let G_o be a compact Lie group with complexification G. Let X be the full flag variety of G. Take $K = G$ in theorem 7.1. Then the theory of $(\mathcal{D}_\lambda, G)$-modules on X reduces to the Borel-Weil-Bott theory of G-homogeneous sheaves of $\mathcal{O}$-modules on X.

Next, one can easily classify the irreducible $(\mathcal{D}, K)$-modules for $SL_2(\mathbf{R})$, and compare the result with the classification of irreducible representations of $SL_2(\mathbf{R})$ in [Vogan, ch. 1]. Here $K = \mathbf{C}^*$, and the flag variety X is the Riemann sphere $\mathbf{P}^1(\mathbf{C})$. There are three orbits $\{0\}$, $\{\infty\}$ and $\mathbf{C}^*$, with respective stabilizers K, K, and $\{\pm 1\}$.

An example to not neglect, is the case of an abelian group: $\mathbf{R}^*$, $U(1)$, or $\mathbf{C}^*$; one can also take the product of several copies of these groups. Then we have $G = \mathbf{C}^*$ in the first two cases, and $G = \mathbf{C}^* \times \mathbf{C}^*$ in the third one. K is equal to $\{\pm 1\}$, G, or diag$\mathbf{C}^*$ respectively, and the flag variety is a point in all three cases.

Building on this experience, one can classify the irreducible representations of $SL_2(\mathbf{C})$. Here $G = SL_2(\mathbf{C}) \times SL_2(\mathbf{C})$, $K = \mathrm{diag} SL_2(\mathbf{C})$, and the full flag variety is $\mathbf{P}^1(\mathbf{C}) \times \mathbf{P}^1(\mathbf{C})$ on which K has two orbits corresponding to the Bruhat cells.

One further easy example is to look at $G = SL_3(\mathbf{C})$, $K = SO_3(\mathbf{C})$, and X is the partial flag variety $\mathbf{P}^2(\mathbf{C})$ asssociated to a maximal parabolic subgroup. Then K has two orbits: $\mathbf{P}^1\mathbf{C}$ imbedded as a quadric, and its complement. To classify the irreducible representations of $SL_3(\mathbf{R})$, we consider the same situation, except that X is now the full flag variety. Then K has four orbits: $Y = \mathbf{P}^1(\mathbf{C})$ imbedded as a quadric, two orbits of dimension two isomorphic to a fibre product of $\mathbf{P}^1$ and the affine line $\mathbf{A}^1$ – they both contain Y in their closures, and one orbit of dimension three which is the complement of the first three.

Finally, a more challenging example is to work out the representation of $U(n, 1)$ by the theory of $\mathcal{D}$-modules. Here $G = GL_{n+1}(\mathbf{C})$, $K = GL_n(\mathbf{C}) \times GL_n(\mathbf{C})$, and X is the variety of complete flags in $\mathbf{C}^{n+1}$. Using elementary enumerative geometry, one checks that K has $n+1$ closed orbits – which are full flag varieties for K, hence have dimension $\frac{n^2-n}{2}$, n orbits of dimension one more – each containing two closed orbits in its closure, $n-1$ orbits of dimension on more – each containing the closure of two previous orbits, and so on until we reach the unique open orbit. Thus the diagram of the closure relations of orbits is a binary tree of depth $n+1$.

II Spherical $\mathcal{D}$-modules

II.1 K-orbits in a flag space.

Let G be a complex connected reductive linear algebraic group. Let $\mathcal{B}$ be the variety of Borel subgroups of G, and $\mathcal{P}$ the variety of subgroups of G conjugate to a fixed parabolic subgroup P. By analogy with the case $G = GL(n)$, we will also refer to $\mathcal{B}$ as the variety of complete flags, or the full flag variety, and $\mathcal{P}$ will be called a partial flag variety. Let K be an algebraic subgroup of G; it acts on $\mathcal{B}$ and $\mathcal{P}$. Recall that K is called admissible if the number of K-orbits in $\mathcal{B}$ is finite. In this case, there is a Zariski open K-orbit which is automatically unique and dense. Conversely:

1.1 Lemma:[Brion] *If an algebraic subgroup of G acts on the full flag variety $\mathcal{B}$ with an open orbit, then it has only a finite number of orbits.*

This property is unfortunately false for partial flag spaces as shown in example 1.3 below. Two subgroups K and B of G are said to be transversal if $\mathbf{k} + \mathbf{b} = \mathbf{g}$. The existence of an open K-orbit on $\mathcal{B}$ is equivalent to the existence of a Borel subgroup B transversal to K. We have a G-equivariant fibration $\pi : \mathcal{B} \to \mathcal{P}$ which assigns to a Borel subgroup B the parabolic subgroup $P \in \mathcal{P}$ containing B. Therefore the finiteness of the number of K-orbits in $\mathcal{B}$ implies this finiteness in $\mathcal{P}$. The fiber of π over P is the variety of Borel subgroups of P. Note that if Y is the closure of one K-orbit in $\mathcal{P}$, then $\pi^{-1}Y$ is a closed K-stable subset of $\mathcal{B}$ with the same number of components as Y hence, it is the closure of one K-orbit in $\mathcal{B}$.

Let θ be an involution of G and put $K = G^\theta$. Then any Iwasawa decomposition of $\mathbf{g}$ with respect to $\mathbf{k}$ shows that K acts with finitely many orbits on $\mathcal{B}$.

1.2 Lemma: *The orbits of K are affinely imbedded in $\mathcal{B}$.*

Proof: Consider the map $h : \mathcal{B} \to \mathcal{B} \times \mathcal{B} : B \to (B, \theta B)$. The diagonal action of G on $\mathcal{B} \times \mathcal{B}$ decomposes this variety into $\#W$ orbits of G where W is the Weyl group of G. To $w \in W$ corresponds the G-orbit C_w of (B, wB) where B is any Borel subgroup of G.

Claim 1: The G-orbits C_w is affinely imbedded in $\mathcal{B} \times \mathcal{B}$ for any $w \in W$.

Indeed, let us consider the projections p_1 and p_2 $: \mathcal{B} \times \mathcal{B} \to \mathcal{B}$ on each factor. For simplicity we fix a Borel subgroup B of G. The B-orbits in $\mathcal{B}$ are called the Bruhat cells: they are indexed by W and we denote by B_w the B-orbit of wB. B_w is isomorphic to an affine space of dimension $\ell(w)$. Let w_o be the longest element of W and $\mathcal{B}^o$ the corresponding cell. Then $\mathcal{B}^o$ is open in $\mathcal{B}$.

Consider the open subset $V = p_1^{-1}(\mathcal{B}^o) \cap p_2^{-1}(\mathcal{B}^o) \subset \mathcal{B} \times \mathcal{B}$; it is isomorphic to $\mathbf{A}^{2\ell(w_o)}$. Since $\mathcal{B} = \bigcup_{w \in W} w\mathcal{B}^o$, the subsets $V_{w_1,w_2} = p_1^{-1}(w_1\mathcal{B}^o) \cap p_2^{-1}(w_2\mathcal{B}^o)$, for $w_1, w_2 \in W$, form an open cover of $\mathcal{B} \times \mathcal{B}$ by affine spaces. It suffices to check that $C_w \cap V_{w_1,w_2}$ is affine for all $w_1, w_1, w_2 \in W$. Consider the map $p_1 : C_w \cap p_1^{-1}(\mathcal{B}^o) \to \mathcal{B}^o \simeq \mathbf{A}^{\ell(w_o)}$. It is surjective since $C_w = G \cdot (B, wB)$ and the fiber over $w_oB \in \mathcal{B}$ is $\{(w_oB, w_obwB) \mid b \in B\} \simeq B_w \simeq \mathbf{A}^{\ell(w)}$. Since it is B equivariant, it is a fibration over an affine space with affine fibers. Restricting $\mathcal{B}^o$ to a smaller open subset U of $\mathcal{B}$, if necessary, we see that $C_w \cap p_1^{-1}(U)$ is affine. Now the complement of $\mathcal{B}^o$ in $\mathcal{B}$ is the set of orbits of non-maximal dimension; it is a connected hypersurface H of $\mathcal{B}$. Hence $C_w \cap p_1^{-1}(U) \cap p_2^{-1}(\mathcal{B}^o)$ is the subset of $\mathcal{B} \times \mathcal{B}$ obtain from $C_w \cap p_1^{-1}(U)$ by removing the points of $\mathcal{B} \times H$. But $C_w \cap p_1^{-1}(U) \cap (\mathcal{B} \times H)$ is a connected hypersurface in $C_w \cap p_1^{-1}(U)$. Hence $C_w \cap p_1^{-1}(U) \cap p_1^{-1}(\mathcal{B}^o)$ is still affine. A similar argument applies to the other open sets V_{w_1,w_2}.

We continue the proof of the lemma.

Every K-orbit in $\mathcal{B}$ is mapped by h into a single G-orbit. Consider one G-orbit Y in $\mathcal{B} \times \mathcal{B}$, $h^{-1}Y$ is a K-stable subset of $\mathcal{B}$, which may be disconnected.

Claim 2: The K-orbits in $h^{-1}Y$ are all open subsets of $h^{-1}Y$.

$$\begin{array}{ccc} h^{-1}Y & \longrightarrow & Y \\ i \downarrow & & \downarrow j \\ \mathcal{B} & \xrightarrow{h} & \mathcal{B} \times \mathcal{B} \end{array}$$

This is a Cartesian square and j is an affine morphism by Claim 1. Since base change preserves affinity, i is also an affine morphism. Claim 2 says that a K-orbit in $h^{-1}Y$ is a union of connected components of $h^{-1}Y$. Hence it is affinely embedded in $\mathcal{B}$.

To prove Claim 2, consider two points B and B' in $h^{-1}Y$ which are close to each other, so that we may write $B' = \exp x\, B$ for some $x \in \mathfrak{g}$. Then $h(B') = (\exp x\, B, \theta(\exp x \cdot B))$ and $h(B) = (B, \theta B)$. Since $h(B)$ and $h(B)$ belong to the same G orbit Y, for B and B' close enough, there exists $y \in \mathfrak{g}$

such that $(\exp xB, \theta(\exp xB)) = (\exp yB, \exp y\theta B)$. But a Borel subgroup is its own normalizer, hence $\exp(-x)\exp y \in B$ and $\exp(-x)\exp\theta y \in B$. Since x and y are small, $\exp(-x)\exp y \approx \exp(-x+y)$ and $\exp(-x)\exp\theta y \approx \exp(-x+\theta y)$. Hence $x - y \in \mathbf{b}$ and $x - \theta y \in \mathbf{b}$. Therefore $x - \frac{y+\theta y}{2} \in \mathbf{b}$ and this implies

$$B' = \exp\left(\frac{y+\theta y}{2}\right) B$$

Obviously $\exp(\frac{y+\theta y}{2}) \in K$. Thus B' is in the same K-orbit as B. This proves the claim and finishes the proof of the Lemma 1.2. □

Now let us consider K-orbits on a general flag space $\mathcal{P}$. The finiteness of the number of orbits in $\mathcal{P}$ is ensured by the fact that K acts with finitely many orbits on $\mathcal{B}$, and the fibration $\mathcal{B} \to \mathcal{P}$ is G-equivariant. But these K-orbits need no longer be affinely embedded.

1.3 Example:

$$G = Gl_3 \qquad P = \begin{pmatrix} * & * & * \\ 0 & * & * \\ 0 & * & * \end{pmatrix} \qquad \mathcal{P} = \mathbf{P}^2(\mathbf{C})$$

$\theta = \mathrm{Ad}\,\mathrm{diag}(1,-1,-1)$ so that $K = Gl_1 \times Gl_2 = \begin{pmatrix} * & 0 & 0 \\ 0 & * & * \\ 0 & * & * \end{pmatrix}$. Then K has three orbits on $\mathcal{P}$: the point 0 corresponding to the line $[x,0,0]$ in $\mathbf{C}^3$, the projective line $H \simeq \mathbf{P}^1$ consisting of the lines in $\mathbf{C}^3$ contained in $\{[0,x_1,x_2] \mid x_1, x_2, \in \mathbf{C}\}$, and the complement C of these two closed orbits. $C = \mathbf{P}^2 \setminus \{\mathbf{P}^1 \cup 0\} = \mathbf{A}^2 \setminus \{0\}$ is not an affine variety. Note that P has two orbits on $\mathcal{P}$: $\{0\}$ and $\mathbf{P}^2 \setminus \{0\}$. The latter is again not affine.

Let us also give a counter-example to lemma 1.1 for partial flag spaces. Consider the subgroup $N = \begin{pmatrix} 1 & 0 & 0 \\ * & 1 & 0 \\ * & 0 & 1 \end{pmatrix}$ of G. The orbit of the point $[1,0,0]$ is open in $\mathcal{P}$ since N acts as the translation action of $\mathbf{A}^2$ on the 'finite' part of $\mathbf{P}^2(\mathbf{C})$. But the action of N on the projective line at infinity H is trivial. Thus N has one open orbit and infinitely many closed orbits in $\mathcal{P}$. However, this example does contradict lemma 1.1, because in the full flag space $\mathcal{B}$ of G, the group N does not have any open orbit; indeed, $\dim N = 2$ while $\dim \mathcal{B} = 3$.

If we analyze the proof of Lemma 1.2 we first have to replace $\mathcal{B} \times \mathcal{B}$ by $\mathcal{P} \times^{\theta} \mathcal{P}$ where ${}^{\theta}\mathcal{P}$ denotes the set of parabolic subgroups of G conjugate to ${}^{\theta}P$. Note that although there always exist θ stable Borel subgroups, θP need not be conjugate to P. (Take for example GL_3 and $K = 0_3$). The G orbits on $\mathcal{P} \times^{\theta} \mathcal{P}$ are parametrized by W/W_L where L is the Levi factor of P and so are the B-orbits on $\mathcal{P}$.

$$\mathcal{P} \times^{\theta} \mathcal{P} = \coprod_{w \in W/W_L} G \cdot (P, \theta(wP)) \quad , \quad \mathcal{P} = \coprod_{w \in W/W_L} BwP$$

The B-orbits on $\mathcal{P}$ are affine spaces. Let $Bw_oP = \mathcal{P}^o$ be the big cell and $\{C_w^P = G \cdot (P, \theta(wP))\}$. Then $C_w^P \cap p_1^{-1}(\mathcal{P}^o) \to \mathcal{P}^o$ is a fibration but the fiber over w_oP is isomorphic to $P \cdot \theta(wP)$, i.e. a P-orbit on ${}^{\theta}\mathcal{P}$. The trouble is that the P orbits on $\mathcal{P}$ or ${}^{\theta}\mathcal{P}$ need not be affinely embedded, *cf.* example above. The second part of the proof of Lemma 1.2 extends easily to the general case.

The fact that a K-orbit Y is affinely embedded means essentially that the boundary ∂Y of Y has codimension 1 in $\overline{Y}$. However, a closed subset of an algebraic variety is always affinely imbedded. Hence the closed K-orbits in $\mathcal{P}$ are affinely imbedded and smooth. These are the orbits with which we shall be concerned in the sequel. The K-orbit of a Borel subgroup B is closed if and only if B is θ-stable, [Matsuki, Springer]. Similarly:

1.4 Lemma: *The K-orbit of* P *in $\mathcal{P}$ is closed if only if* P *contains a θ-stable Borel subgroup.*

Proof: Let $Y = K \cdot P$ and $\pi : \mathcal{B} \to \mathcal{P}$. If Y is closed then $\pi^{-1}Y$ is the closure of one K-orbit. Hence $\pi^{-1}Y$ contains a closed K-orbit say $K \cdot B$ in $\mathcal{B}$. By Matsuki-Springer's characterization, B is θ-stable and since $\pi(B) \in Y$, there is a K-conjugate of P which contains B. Hence P contains a θ stable Borel subgroup.

Conversely, if P contains a θ-stable Borel subgroup B, then $Y = \pi(K \cdot B)$. Again by Matsuki-Springer, $K \cdot B$ is closed, hence compact. Therefore Y is compact. □

Note that if $K \cdot P$ is closed, then $K \cap P$ is parabolic in K and $K \cdot P$ is isomorphic to the flag space of K of type $P \cap K$.

The map $h \cdot \mathcal{B} \to \mathcal{B} \times \mathcal{B} : B \mapsto (B, \theta B)$ can easily be related to Springer's parametrization of K-orbits on $\mathcal{B}$. Choose a θ-stable Borel subgroup B and a θ-stable cartan subgroup T in B. Let $A = \{g \in G \mid g^{-1}\theta g \in N_G(T)\}$. K acts on the left of A and T on the right. Put $V = K \setminus A/T$.

1.5 Proposition:(Springer) $\mathcal{B} = \coprod_{v \in V} K \cdot vB$

On the other hand, $\mathcal{B} \times \mathcal{B} = \coprod_{w \in W} G \cdot (B \cdot wB)$. Since h maps K-orbits into G-orbits, it induces a map $\bar{h} : V \to W : n \mapsto n^{-1}\theta n$. The image of $\bar{h}$ consists of θ-twisted involutions, i.e. elements $w \in W$ such that $w \cdot \theta(w) = 1$.

II.2 $(\mathcal{D}, K)$-modules with a K-fixed vector

Let $\mathbf{g} = \mathbf{k} \oplus \mathbf{a} \oplus \mathbf{n}$ be an Iwasawa decomposition of $\mathbf{g}$ with respect to θ. It is not unique but all choices are conjugate by K. At the group level KAN is only an open dense subset of G. Let $L = \mathrm{Cent}_G(A)$ and $P = LN$; this is a Levi decomposition of the parabolic subgroup P. The G-conjugacy class of P is uniquely determined by θ, and we say that P is associated to K. The corresponding flag variety $X = \mathcal{P}$ is also called associated to K and $K \cdot P$ is the open K orbit in X.

Note that if $G(\theta, R)$ denotes a real form of G whose Cartan involution is θ and if $\mathbf{a}$ is chosen to be defined over R, then P is the complexification of a minimal parabolic subgroup of $G(\theta, R)$.

An $(\mathbf{g}, K)$-module V is called K*-spherical* if the space V^K of K-invariant vectors in V is nonzero. As will be explained in the next section, the irreducible K-spherical $(\mathbf{g}, K)$-modules are those which are realizable as functions over the real symmetric space $G(\theta, R)/K(R)$. Kostant classified the irreducible K-spherical representations of $G(\theta, R)$: they all are quotients of principal series representations induced from one-dimensional representations $e^{\lambda} \otimes 1_N$, $\lambda \in \mathbf{t}_p$, of a minimal parabolic subgroup $P(\theta, \mathbf{R}) = L(\theta, R)N(\theta, R)$ of $G(\theta, R)$ *cf.* [Kostant]. Let $\mathcal{D}$ be a tdo on X.

2.1 Definition: We shall say that a $(\mathcal{D}, K)$-module $\mathcal{M}$ on X is K*-spherical* if $\mathrm{Hom}_{(\mathcal{O},K)}(\mathcal{O}, \mathcal{M}) \neq 0$. This notion is interesting mainly for simple or standard $(\mathcal{D}, K)$-modules. We also say that $\mathcal{M}$ has *trivial K-isotropy* if for every K-orbit Y in the support of $\mathcal{M}$, the isotropy group K_y acts trivially on the fiber of $\mathcal{M}$ at $y \in Y$.

A $(\mathcal{D}_\lambda, K)$-module $\mathcal{M}$ is called *standard* if it is the maximal extension $i_*\mathcal{L}$ of an invertible K-sheaf $\mathcal{L}$ on an affinely embedded K-orbit $i : Y \to X_i$, it is called *costandard* if it is the minimal extension $i_!\mathcal{L}$, and $\mathcal{M}$ is *simple* if it is the middle extension $i_{!*}\mathcal{L}$. Of course $\mathcal{L}$ corresponds to a representation $\tau : K_x \to \mathbf{C}^\times$, $y \in Y$, such that $d\tau$ coincides with $\lambda - \rho_p$ on $\mathbf{k}_y \cap \mathbf{t}_p$ so we will often denote $\mathcal{L}$ by $\mathcal{O}_Y(\lambda,\tau)$. We have also the sequence $i_!\mathcal{L} \to i_*\mathcal{L} \mapsto i_*\mathcal{L}$.

2.2 Theorem: *Let $\mathcal{M}$ be a simple or a standard $(\mathcal{D},K)$ module on X. Then $\mathcal{M}$ is K-spherical if and only if $\operatorname{supp}\mathcal{M} = X$ and $\mathcal{M}$ has trivial isotropy. Moreover $\mathcal{M}$ has at most one K-invariant section, up to scalars.*

Proof: $\mathrm{Hom}_{(\mathcal{O},K)}(\mathcal{O},\mathcal{M})$ is a vector space and we want to compute its dimension. One can verify that the only K-orbit on a flag space which can support a standard $(\mathcal{D},\mathcal{K})$-module containing the trivial representation of K is the open K-orbit. The sections of $\mathcal{O}$ are determined by their restriction to the open K-orbit X^0. Let $i : x \hookrightarrow X^0$ be the inclusion of a point in X^0. By K-equivariance, it suffices to compute $\dim \mathrm{Hom}_{K_x}(\mathbb{1}, i^!\mathcal{M})$ where K_x is the stabilizer of x in K and $\mathbb{1}$ is the trivial module $\mathbf{C}$. This expression shows that if $\mathcal{M}$ is K-spherical, then $x \in \operatorname{supp}\mathcal{M}$, and by K-equivariance $X^0 \subseteq \operatorname{supp}\mathcal{M}$, hence $X = \operatorname{supp}\mathcal{M}$. Let $j : X^0 \hookrightarrow X$. Since $\mathcal{M}$ is simple or standard it can be written as $j_!\mathcal{L}, j_*\mathcal{L}$ or $j_{!*}\mathcal{L}$ for some line bundle $\mathcal{L}$ on X^0 corresponding to a representation τ of K_x. Now $i^!\mathcal{M} = i^!\mathcal{L} = \mathbf{C}_\tau$. Hence

$$\begin{aligned} \dim \mathrm{Hom}(\mathbf{1}, \mathbf{C}_\tau) &= 1 \quad \text{if } \tau \text{ is trivial} \\ &= 0 \quad \text{otherwise.} \end{aligned}$$

The converse assertion is clear from the above discussion. □

2.3 Remark: [Harish Chandra] has proved that if δ is an irreducible representation of K and V is an irreducible quasisimple Banach space representation of $G(\theta, R)$, then

$$\mathrm{mtp}(\delta, V) \leq \dim(\delta)$$

cf. [Godement]. In particular if $\delta = \mathbf{1}$ then this implies $\mathrm{mtp}(\mathbf{1}, V) \leq 1$, as in the above proposition.

II.3 Relations with the analytic theory.

It is inspiring to bear in mind the relations between the $\mathcal{D}$-module picture and the analytic picture on symmetric spaces. Let G_o denote a real reductive Lie group obtained as follows. Consider the complex connected reductive algebraic group G and let θ be an involution of G with fixed point group K. Take $G_o = G(\theta, \mathbf{R})$ to be a real form of G such that θ is a Cartan involution of G_o, i.e. $K_o = K(\theta, \mathbf{R}) = K \cap G_o$ is a maximal compact subgroup of G_o.

Consider another involution σ of G which commutes with θ. Let H be the fixed point set of σ in G and let H_o be $H \cap G_o$. Using σ we can also define another real form of G, namely $G^r := G(\sigma, \mathbf{R})$ and $K^r := K \cap G^r$. $H^r := H \cap G^r$. Observe that by definition H^r is a maximal compact subgroup of G^r. The symmetric space G^r/H^r is Riemannian, and it is called the dual of G_o/H_o. Note that exchanging the involutions θ and σ amounts to exchanging the subscript o for a superscript r. The set-up with superscript r is called the Riemannian dual of the situation with subscript o.

Consider an Iwasawa decomposition of $\mathbf{g}$ with respect to $\sigma : \mathbf{g} = \mathbf{h} \oplus \mathbf{a} \oplus \mathbf{n}$, and let $W(\mathbf{a})$ be the Weyl group of $\mathbf{a}$ in $\mathbf{g}$. Let $\mathbf{D}(G_o/H_o)$ and $\mathbf{D}(H^r/G^r)$ be the algebras of invariant differential operators on G_o/H_o and G^r/H^r respectively. They are naturally via holomorphic differential operators on G/H and there is an isomorphism $\mathbf{D}(G/H) \simeq S(\mathbf{a})^{W(\mathbf{a})}$, or if you prefer at the level of maximal spectra, we have: $\mathbf{a}^*/W(\mathbf{a}) \simeq \operatorname{Max} \mathbf{D}(G/H)$ where a shift by ρ_p is included. In particular these algebras are commutative and every $\lambda \in \mathbf{a}^*$ defines a character χ_λ of both algebras.

We are going to study the irreducible representations V of G_o which occur in $C^\infty(G_o/H_o)$; as is customary we assume that V is quasi-simple, i.e. the center Z of U acts by scalars on V. Note that Z is the algebra of left and right G-invariant differential operators on G, hence there is a natural map $Z \to \mathbf{D}(G/H)$. We shall see in Section IV.3 that this map is not always surjective. However, in this chapter we shall only study the irreducible representations of G_o which occur within one eigenspace $C^\infty(G_o/H_o, \chi_\lambda) =$ the χ_λ-eigenspace of $\mathbf{D}(G_o/H_o)$ in $C^\infty(G_o/H_o)$.

Next by a result of [Casselman, p. 44] and [Wallach, p. 320], there is a natural C^∞ topology on any irreducible Harish-Chandra module coming from a canonical globalization of M which is an irreducible smooth representation M^∞ of G_o. In particular, one can deduce from their result that every K-invariant functional on M extends continuously to M^∞, i.e. $\operatorname{Hom}_K(M, \mathbb{1}) = \operatorname{Hom}_{K_o}(M^\infty, \mathbb{1})$, where the second Hom contains only continuous functionals, see [van den Ban – Delorme] for details. This result implies that the study of the irreducible G_o-submodules of $C^\infty(G_o/H_o)$ is equivalent to the study of the irreducible $(\mathbf{g}, K)$-submodules of $C^\infty{}_K(G_o/H_o) :=$ the space of K_o-finite differentiable functions on G_o/H_o.

Assembling these two observations, the problem is reduced to the decomposition of the $(\mathbf{g}, K)$-modules of $C^\infty(G_o/H_o; \chi_\lambda)$. Because of the K_o-finiteness, these eigenfunctions are automatically real analytic. Thus, the object of interest is really the space $A_K(G_o/H_o;\chi_\lambda)$ of real analytic K_o-finite functions on G_o/H_o which are eigenfunctions of $\mathbf{D}(G/H)$ for the eigencharacter χ_λ.

3.1 Lemma: *Let V be an irreducible quasi-simple representation of G_o. Then the multiplicity of V as submodule of $C^\infty(G_o/H_o)$ is finite.*

Observe that this multiplicity is the dimension of

$$\mathrm{Hom}_{G_o}(V, C^\infty(G_o/H_o)) = \mathrm{Hom}_{H_o}(V, \mathbb{1})$$

by Fröbenius reciprocity. Let V_H be the space of H_o-coinvariants of V. $V_H = \mathbf{1} \otimes_{H_o} V$, $V/\mathbf{h}V$ surjects onto V_H, and $(V_H)^* = \mathrm{Hom}_{H_o}(V, \mathbb{1}) = (V^*)^H$. By the above result of Casselman-Wallach-van den Ban-Delorme, $V/\mathbf{h}V = V^0/\mathbf{h}V^0$ where V^0 is the Harish-Chandra module of V. $V^0/\mathbf{h}V^0$ can be considered as a space of functions on G_o which are eigenvectors of Z, left K_o-finite and right invariant by H_o. These conditions form a holonomic system of differential equations with regular singularities. It is known that its solution space is finite dimensional.

Let us mention that the multiplicity of V in $C^\infty(G_o/H_o)$ as submodule i.e. $\dim V_H$ can be quite smaller than the multiplicity of V has subquotient. However, they are both finite and thanks to the following result we have an upper bound on the multiplicity of V as subquotient. Let W be the complex Weyl group of $\mathbf{g}$. Let $\mathbf{g} = \mathbf{h} \oplus \mathbf{r} = \mathbf{k} \oplus \mathbf{s}$; let $\mathbf{a}_1$ be a maximal abelian subspace of $\mathbf{r} \cap \mathbf{s}$ and $\ell = \mathrm{cent}(\mathbf{a}_1; \mathbf{g})$. Let W_ℓ be the complex Weyl group of ℓ.

3.2 Proposition: [van den Ban] *Let δ be an irreducible – hence finite dimensional – representation of K_o, and $\lambda \in a^*$. Then:*

$$\mathrm{mtp}(\delta; A_K(G_o/H_o;\chi_\lambda)) \leq \dim(\delta)\, \# W/W_\ell$$

The big advantage of Riemannian symmetric spaces is the Helgason isomorphism between $A(G^r/H^r;\chi_\lambda)$ and a space $B(X^r; L_\lambda)$ of hyperfunctions. More precisely, consider the minimal parabolic subgroup P^r in G^r defined by $P^r = \mathrm{Cent}_{G^r}(A^r)\cdot N_o$ where A^r, N^r correspond to an Iwasawa decomposition of $G^r = H^r A^r N^r$. Then put X^r = the variety of parabolic subgroup of G^r conjugate to P^r. As in Chapter 1, we can work invariantly at every point

P^r of X^r, and define Cartan factors $T_P^r = P^r/(P^r, P^r)$ which are canonically conjugate for different points $P^r \in X^r$. The θ-split component of T_P^r is $A^r \simeq (R_+^*)^\ell$ where ℓ is the split rank of G^r and $\mathrm{Hom}(A^r, \mathbf{C}^\times) \simeq \mathbf{a}^*$. Every $\lambda \in \mathbf{a}^*$ determines a line bundle L_λ over X^r corresponding to the representation $\lambda - \rho_p$ of P^r. Then $B(X^r; L_\lambda)$ is the space of hyperfunctions sections of L_λ over X^r.

The Poisson transform is

$$\begin{aligned} S : B(X^r; L_\lambda) &\longrightarrow A(G^r/H^r; \chi_\lambda) \\ f &\longmapsto \textstyle\int_{K^r} f(\cdot k)dk \,. \end{aligned}$$

It will be explained in more details in section 8. Here we have identified $B(X^r, L_\lambda)$ with hyperfunctions on G^r which transform according to $\lambda - \rho_p$ for the right action of P^r. Helgason proved that S is an isomorphism between K-finite vectors and he conjectured that S is an isomorphism of topological spaces. This was settled positively by [Kashiwara, Kowata, Minemura, Okamoto, Oshima and Tanaka].

One should observe that $B(X^r, L_\lambda)$ depends on $\lambda \in \mathbf{a}^*$, while $A(G^r/H^r; L_\lambda)$ depends only on the orbit $W(\mathbf{a}) \cdot \lambda$, and the definition of S is independent of λ. In some sense, S has several candidates for its inverse map $\beta_{w\lambda}$: $A(G^r/H^r; \chi_\lambda) \to B(X_o; L_{w\lambda})$ for every $w\lambda \in W(\mathbf{a})\lambda$. Here $\beta_{w\lambda}$ is a boundary value map, because X^r can be viewed as a piece of the boundary of G^r/K^r, in the direction w, see [Oshima–Matsuki, p. 346]. If λ is dominant, then $S \circ \beta_\lambda = 1$ and $\beta_\lambda \circ S = 1$, but if λ is not dominant, β_λ may have fail to be injective, and hence surjective.

One of the most useful tools in the study of indefinite symmetric spaces is the Flensted-Jensen isomorphism:

$$\eta : \mathcal{A}_K(G_o/H_o; \chi_\lambda) \xrightarrow{\sim} \mathcal{A}_K(G^r/H^r; \chi_\lambda)$$

which is obtained by analytic continuation to the complex symmetric space G/H. The right-hand side is the space of real analytic functions on the Riemannian space G^r/H^r which are eigenfunctions of $\mathbf{D}(G^r/H^r) \simeq S(\mathbf{a})^{W(\mathbf{a})}$ for the same eigencharacter χ_λ, and which are K^r-algebraic, i.e. they transform under K^r according to the restriction of an algebraic representation of K. Since the algebraic representations of K^r are in natural bijection with the representations of K_o, via analytic continuation to the complex group

G, our notation is not ambiguous. This isomorphism is generally stated for connected groups but one can extend its proof in [Schlichtkrull] to our set-up.

In the following sections of this chapter, we give a description of the $(\mathbf{g}, K)$ -submodules of $A_K(G_o/H_o; \chi_\lambda)$ in terms of $(\mathcal{D}_\lambda, K)$-modules on the complex flag variety $X = G/P$, where P is now the complexification of a maximal parabolic subgroup P^r of G^r. However one should understand that this project contains objects of two different natures. When we speak of a $(\mathbf{g}, K)$-submodule V of $A_K(G_o/H_o; \chi_\lambda)$, we are given a concrete realization of V. Hence we should focus on $(\mathcal{D}_\lambda, K)$-modules $\mathcal{M}$ with a concrete realization on X , for example the standard ones.

If we do not want to specify a concrete realization, we can work with H-spherical objects. But, H does not act on a $(\mathbf{g}, K)$ or $(\mathcal{D}, K)$-module. So we will have to use a functor which transforms $(\mathbf{g}, K)$-modules into $(\mathbf{g}, H)$-modules. The construction of the analogous functor for $\mathcal{D}$-modules will be explained in §4, and used in §5.

On the other hand, $X^r := G^r/P^r$ is a real algebraic subvariety of X . If we have a standard H-spherical $(\mathcal{D}, K)$-module on X, we would like to compare it with the space of hyperfunctions along X^r . This question was raised by Flensted-Jensen, and we will answer it in §6.

We can summarize these relations by a diagram:

$$
\begin{array}{ccc}
\begin{array}{c} H\text{-spherical} \\ (\mathcal{U}_{\chi_\lambda}, K)\text{-modules} \end{array} & \underset{\Gamma}{\overset{\Delta_\lambda}{\rightleftarrows}} & \begin{array}{c} H\text{-spherical} \\ (\mathcal{D}_\lambda, K)\text{-modules on } X \end{array} \\
\updownarrow & & \updownarrow \\
A_K(G_o/H_o; \chi_\lambda) & \underset{\eta \circ S}{\overset{\beta_\lambda \circ \eta}{\rightleftarrows}} & B_{K^{r\prime}}(X^r; L_\lambda)
\end{array}
$$

The horizontal arrows are bijective when λ is dominant, regular. The vertical arrows involve concrete realizations, and are not defined for all H-spherical modules. The case of square integrable functions and closed K-orbits in X will be studied in §7.

The Poisson transform cannot be defined directly for an indefinite symmetric space G_o/H_o, because H_o is not compact. We will see in §8, that one can define a map between $(\mathcal{D}_\lambda, K)$-modules on X and $(\mathcal{D}_{\chi_\lambda}, K)$-modules on G/H.

3.3 Remark: There is a duality involved in the passage from H-spherical representations of G_o to submodules of $C^\infty(G_o/H_o)$. V is an irreducible submodule of $C^\infty(G_o/H_o)$ if and only if $V_H \neq 0$, *i.e.* if and only if $(V^*)^H \neq 0$ where V^* is the continuous dual of V, but V^H may well be zero. However if M is a $(\mathbf{g}, H)$-module, then H acts algebraically on M and it is clear that

$$M_H \neq 0 \Longleftrightarrow M^H \neq 0 .$$

We will see from the classification of H-spherical $(\mathcal{D}, K)$ modules that for a representation V of G_o

$$V_H \neq 0 \Longleftrightarrow (\tilde{V})_H \neq 0$$

where $\tilde{V}$ is the contragredient representation. In this sense, H carries well the name of symmetric subgroup. At the other extreme, there is N_o : a maximal unipotent subgroup of G_o. Then

$$V_N \neq 0 \Longleftrightarrow (\tilde{V})_{\overline{N}} \neq 0$$

where $\overline{N}$ is the subgroup opposite (or transposed) to N.

II.4 Going from $(\mathcal{D}, K)$ modules to $(\mathcal{D}, H)$ modules

The new results of this section are due to J. Bernstein.

If B is a subgroup of H, (we will take $B = H \cap K$ for the applications), one can construct two functors Γ_B^H and L_B^H from the category of $(\mathbf{g}, B)$-modules to the category of $(\mathbf{g}, H)$-modules. The functor Γ was introduced by Zuckerman who showed that the derived functors of Γ are quite meaningful for representation theory, see [Vogan, 1 p. 325]. The functor L appears in [Enright-Wallach], where it is defined using a double duality. It is the inductive version of Γ in the following sense.

Given two abelian categories one can define the notion of inductive functors and projective functors. The inductive functors are those which are right exact, commute with products, and with (inverse) limits, so they can be left adjoint. The projective functors are those which are left exact, commute with coproducts (sums), and with colimits (direct limits), so they can be right adjoint. For example, given a parabolic subalgebra $\mathbf{p}$ of $\mathbf{g}$ with Levi decomposition $\mathbf{p} = \ell \oplus \mathbf{n}$ and $B = H \cap L$, we may construct the following functors. The notation is as follows: if A is an algebra or a group, A-mod denotes the category of A-modules. Let $R(H)$ be the ring of regular functions on H. For simplicity, assume $\mathbf{C}_\rho = (\wedge^{\mathrm{top}}\mathbf{n})^{1/2}$ is a representation of B.

$$
\begin{array}{rrcl}
\text{ext:} & (\ell, B)\text{-mod} & \longrightarrow & (\mathbf{p}, B)\text{-mod} \\
 & M & \longrightarrow & M \otimes \mathbf{C}_\rho \\
\mathbf{n}\text{-coinv:} & (\mathbf{p}, B)\text{-mod} & \longrightarrow & (\ell, B)\text{ mod} \\
 & M & \longrightarrow & M_{\mathbf{n}} \otimes \mathbf{C}_\rho := M/\mathbf{n}M \otimes \mathbf{C}_{-\rho} \\
\mathbf{n}\text{-inv:} & (\mathbf{p}, B)\text{-mod} & \longrightarrow & (\ell, B)\text{-mod} \\
 & M & \longrightarrow & M^{\mathbf{n}} \otimes \mathbf{C}_\rho := \{x \in M \mid \mathbf{n}x = 0\} \otimes \mathbf{C}_{-\rho} \\
\text{ind:} & (\mathbf{p}, B)\text{-mod} & \longrightarrow & (\mathbf{g}, B)\text{-mod} \\
 & M & \longrightarrow & U(\mathbf{g}) \underset{U(\mathbf{p})}{\otimes} M \\
\text{pro:} & (\mathbf{p}, B)\text{-mod} & \longrightarrow & (\mathbf{g}, B)\text{-mod} \\
 & M & \longrightarrow & \mathrm{Hom}_{\mathbf{p}}(U(\mathbf{g}), M)_{B-\text{finite}} \\
\text{res:} & (\mathbf{g}, B)\text{-mod} & \longrightarrow & (\mathbf{p}, B)\text{-mod} \\
 & M & \longrightarrow & M \\
\Gamma\text{:} & (\mathbf{g}, B)\text{-mod} & \longrightarrow & (\mathbf{g}, H)\text{-mod} \\
 & M & \longrightarrow & (R(H) \otimes M)^{(\mathbf{h},B)\text{-inv}} \\
L\text{:} & (\mathbf{g}, B)\text{-mod} & \longrightarrow & (\mathbf{g}, H)\text{-mod} \\
 & M & \longrightarrow & (R(H) \otimes M)_{(\mathbf{h},B)\text{-coinv}} \\
F\text{:} & (\mathbf{g}, H)\text{-mod} & \longrightarrow & (\mathbf{g}, B)\text{-mod} \\
 & \bigoplus\limits_{\delta \in \hat{H}} M_\delta & \longrightarrow & \bigoplus_{\delta \in \hat{H}} M_\delta \\
F^\vee\text{:} & (\mathbf{g}, H)\text{-mod} & \longrightarrow & (\mathbf{g}, B)\text{-mod} \\
 & \bigoplus_{\delta \in \hat{H}} M_\delta & \longrightarrow & (\prod\limits_{\delta \in \hat{H}} M_\delta)_{B-\text{finite}}
\end{array}
$$

Note: If H is disconnected, ΓM is the module induced from $B \cdot H^\circ$ to H from the largest $B \cdot H^\circ$ finite submodule of M, similarly for L.

Moreover, all of these categories are endowed with a contragredient functor $\tilde{\cdot}$; for example if $M \in (\mathbf{g}, H)$-mod, then $\widetilde{M} = K$-finite dual of M. The classification goes as follows with adjoint functors facing each other:

inductive functors	**duality relations**	**projective functors**
ext	$\widetilde{M_{\mathbf{n}}} = (\widetilde{M})^{\mathbf{n}}$	$\mathbf{n}$-inv
$\mathbf{n}$-coinv	$\widetilde{\text{ext}\, M} = \text{ext}\, \widetilde{M}$	ext
ind	$\widetilde{\text{ind}\, M} = \text{pro}\, \widetilde{M}$	res

res	$\widetilde{\operatorname{res} M} = \operatorname{res} \widetilde{M}$	pro
L	$\widetilde{LM} = \Gamma \widetilde{M}$	$F^\vee$
F	$\widetilde{FM} = F^\vee \widetilde{M}$	Γ

Now put $d = \dim \mathbf{h}/\mathbf{n}$. Let us denote by L^i the i-left derived functor of L and by Γ^j the jth right derived functor of Γ. To be rigorous L is defined for negative indices and Γ^j is defined for positive indices.

4.1 Proposition: $L^{-i} \simeq \Gamma^{d-i} \otimes \wedge^d(\mathbf{h}/\mathbf{b})$.

The proof is at the end of this section.

When working with $\mathcal{D}$-modules we should use inductive functors, because they will commute with the operation of taking the fiber of a sheaf. If K is not a subgroup of H, we can still define L_K^H by $L_{H\cap K}^H \circ F_K^{H\cap K}$. Let us describe more precisely what is L in our situation. K, H and G are complex reductive groups.

For $M \in (\mathbf{g}, K)$-mod, $L_K^H M$ considered as an $(\mathbf{h}, H)$-module is simply $R(H)\otimes_{(\mathbf{h},K\cap H)} M$. Now to put a structure of $(\mathbf{g}, H)$-module on LM, observe that a $(\mathbf{g}, H)$ module M is simply an $(\mathbf{h}, H)$-module together with an H-map $\mathbf{g}\otimes M \to M$. By definition for any $(\mathbf{g}, H)$-module V we want $L(FV\otimes M) = V \otimes LM$. So we get a map $\mathbf{g} \otimes LM = L(F\mathbf{g} \otimes M) \to LM$ as desired. It is possible to explicit the $(\mathbf{g}, H)$-module structure of LM. Let $R'_H(G)$ be the space of H-algebraic hyperfunctions on G supported on H. In other words consider the inclusion $i : H \to G$ of complex algebraic groups and put $R'_H(G) = \Gamma(G, i_*\mathcal{O}_H) = H_H^c(G, \mathcal{O}_G)$ where $c = \operatorname{codim}(H; G)$, i_* is the direct image in the sense of $\mathcal{D}$-modules on H and G, and $\mathrm{H}_H^c(G, \mathcal{O}_G)$ is a local cohomology group. Then for a $(\mathbf{g}, K)$-module M:

$$L_K^H(M) = R'_H(G) \bigotimes_{(\mathbf{g},K\cap H)} F_K^{K\cap H}(M) .$$

Note to define Γ one can put $\Gamma M = \operatorname{Hom}_{(\mathbf{h},H\cap K)}(R(\mathbf{h}), M)$ and use the same trick as above.

We are going to construct a functor $\mathcal{L}_K^H$ from $(\mathcal{D}, K)$ modules on X to $(\mathcal{D}, H)$ modules on X. As before we first use a forgetful functor from $(\mathcal{D}, K)$-mod to $(\mathcal{D}, K \cap H)$-mod. Then consider the diagram

$$X \xleftarrow{p} H \times X \xrightarrow{q} H \underset{H\cap K}{\times} X \xrightarrow{a} X$$

where p is the projection on the second factor and q is the quotient map given by the diagonal action of $H \cap K$ on the right of H and on X, and a is the action morphism of H on X.

4.2 Lemma: *Let Y be a smooth variety on which the group B acts freely. Put $q : Y \to Y/B$. Then $q^\circ : \mathcal{D}_{Y/B}-mod \longrightarrow (\mathcal{D}_Y, B)-mod$ is an equivalence of categories.*

Recall that q° is the inverse image in the category of $\mathcal{O}$-modules. The inverse functor q_+ consists in taking the B-invariant elements in the direct image: for a $\mathcal{D}_Y$-module $\mathbf{F}$, $q_+\mathcal{F} = (q_\circ\mathcal{F})^B$, where $q_\circ$ is the direct image in the category of sheaves of $\mathcal{O}$-modules.

4.3 Definition: *Let $\mathcal{M}$ be a $(\mathcal{D}, K)$ module on X. Set*

$$\mathcal{L}_K^H\mathcal{M} := a_* q_+ \mathrm{pr}^\circ F_K^{K\cap H}(\mathcal{M})$$

$\mathcal{L}_K^H\mathcal{M}$ is a well-defined $(\mathcal{D}, H)$-module for $\mathcal{D}$ has a G-action. q_+ enters the formula because we want an H-action on $\mathcal{L}_K^H\mathcal{M}$ which is as compatible as possible with the K-action on $\mathcal{M}$. We will only need the case where $H/K \cap H$ is an affine variety, then a_* is well-defined without recourse to derived categories.

4.4 Theorem: $\mathcal{L}_K^H \circ \Delta_\lambda = \Delta_\lambda \circ L_K^H$ *for λ dominant in $\mathbf{t}_P^*$.*

The proof of this result is explained below.

Proof of Proposition 4.1: $L^{-i} \simeq \Gamma^{d-i} \otimes \Lambda^d(\mathbf{k}/\mathbf{b})$. Let M be a $(\mathbf{g}, B)$-module and V a $(\mathbf{g},K)$-module. We have to relate $\mathrm{Hom}_{(\mathbf{g},K)}(V, \Gamma M)$ and $\mathrm{Hom}_{(\mathbf{g},K)}(V, LM)$.

A $(\mathbf{g}, K)$ module V is just a K-module with a map $\mathbf{g} \otimes V \to V$ compatible with the representations of K on both sides. Γ and L commute with tensoring by the finite dimensional representation $\mathbf{g}$ of K. So it suffices to relate the spaces

$$\mathrm{Hom}_K(E, \Gamma M) \qquad \text{and} \qquad \mathrm{Hom}_K(E, LM)$$

where E is now a finite dimensional representation of K. Moreover $\mathrm{Hom}_K(E, \Gamma M) = \mathrm{Hom}_K(\mathbf{1}, E^* \otimes \Gamma M)$ and $\mathrm{Hom}_K(E, LM) = \mathrm{Hom}_K(\mathbb{1}, E^* \otimes LM)$. Again, because Γ and L commute with tensoring by algebraic representations of K, it suffices to relate the spaces.

$$\mathrm{Hom}_K(\mathbf{1}, \Gamma M) \qquad \text{and} \qquad \mathrm{Hom}_K(\mathbf{1}, LM)\,.$$

Now $\mathrm{Hom}_K(\mathbf{1}, \Gamma M) = \mathrm{Hom}_{(\mathbf{k},B)}(\mathbb{1}, M)$ and the right derivatives of this module are the spaces $H^*(k, B; M)$. On the other hand, $\mathrm{Hom}_K(\mathbf{1}, L, M) = \mathbb{1} \otimes_{(\mathbf{k},B)} M$ and the left derivatives of this module are the spaces $H_*(\mathbf{k}, B; M)$.

Let I be the ideal generated by $\mathbf{b}$ in the exterior algebra $\wedge^\bullet \mathbf{k}$ and put $I^i = I \cap \wedge^i \mathbf{k}$. Cohomology is computed using the standard complex $\wedge^\bullet \mathbf{k}/I^\bullet$ with the usual differential, which gives an acyclic $(\mathbf{k}, B)$ resolution of the trivial module $\mathbb{1}$. Then the ith homology group of the complex $\mathrm{Hom}_{(\mathbf{k},B)}(\wedge^\bullet \mathbf{k}/I^\bullet, M)$ is $H^i(\mathbf{k}, B; M)$. Homology is computed using the same standard complex. The ith homology group of the complex $\wedge^\bullet \mathbf{k}/I^\bullet \otimes_{(\mathbf{k},B)} M$ is $H_i(\mathbf{k}, B; M)$.

Let $\mathbf{b}^\perp$ be the orthogonal of $\mathbf{b}$ in $\mathbf{k}^*$; $\mathbf{b}^\perp = (\mathbf{k}/\mathbf{b})^*$. Then

$$\mathrm{Hom}_{(\mathbf{k},B)}(\bigwedge^i \mathbf{k}/I^i, M) \simeq \bigwedge^i \mathbf{b}^\perp \bigotimes_{(\mathbf{k},B)} M .$$

The top degree of $\wedge^\bullet \mathbf{k}/I^\bullet$ is $d = \dim(\mathbf{k}/\mathbf{b})$. Moreover as $(\mathbf{k}, B)$ modules: $\wedge^i \mathbf{k}/I^i \simeq \wedge^i(\mathbf{k}/\mathbf{b})$. Now we have an identification of $(\mathbf{k}, B)$ modules

$$\psi : \wedge^i(\mathbf{k}/\mathbf{b}) \xrightarrow{\sim} \wedge^{d-i}\mathbf{b}^\perp \otimes \wedge^d(\mathbf{k}/\mathbf{b}) .$$

because $\wedge^{d-i}\mathbf{b}^\perp$ is dual to $\wedge^{d-i}(\mathbf{k}/\mathbf{b})$ and we can view $\wedge^i(\mathbf{k}/\mathbf{b})$ as the space of linear maps from $\wedge^{d-i}(\mathbf{k}/\mathbf{b})$ to $\wedge^d(\mathbf{k}/\mathbf{b})$.

Thus we obtain Poincaré duality

$$H_i(\mathbf{k}, B; M) \simeq H^{d-i}(\mathbf{k}, B; M \bigotimes \bigwedge^d(\mathbf{k}/\mathbf{b}))$$

The left-hand side is $L^{-i}\mathrm{Hom}_K(\mathbb{1}, LM)$ while the right-hand side is $R^{d-i}\mathrm{Hom}_K(\mathbb{1}, \Gamma M \otimes \wedge^d(\mathbf{k}/\mathbf{b}))$. This proves the assertion. □

Proof of Theorem 4.4: $\mathcal{L}_K^H \circ \Delta_\lambda = \Delta_\lambda \circ L_K^H$, for $\lambda \in \mathbf{t}_P^*$ dominant. Let $\mathcal{M}$ be a $(\mathcal{D}_\lambda, K)$ module on the flag variety X. Since λ is dominant, it suffices to prove that:

$$\Gamma(X, \mathcal{L}_K^H \mathcal{M}) = L_K^H \Gamma(X, \mathcal{M})$$
$$\mathcal{L}_K^H \mathcal{M} = a_* q_+ p^\circ \mathcal{M}$$
$$X \xleftarrow{p} H \times X \xrightarrow{q} H \underset{K\cap H}{\times} X \xrightarrow{a} X .$$

Let us consider $\mathcal{L}_K^H \mathcal{M}$ has and $(\mathcal{O}, H)$-module first: $\Gamma(H \times X, p^! \mathcal{M}) = R(H) \otimes \Gamma(X, \mathcal{M})$ where $R(H)$ denotes the ring of regular functions on H. Also

$$\Gamma(H \underset{K\cap H}{\times} X, q_+ p^\circ \mathcal{M}) = R(H) \underset{K\cap H}{\otimes} \Gamma(X, \mathcal{M}) .$$

Finally, $\Gamma(X, a_* q_+ p^\circ \mathcal{M}) = R(H) \otimes_{\mathbf{h}, K\cap H} \Gamma(X, \mathcal{M})$. So at the level of H-modules the functor $\mathcal{L}_K^H$ commutes with $\Gamma(X, \cdot)$.

Now we examine the $\mathbf{g}$-action using the functorial properties of L. Built in the definition of L, there is a projection formula

$$L_{K\cap H}^H(V \otimes M) = V \otimes L_{K\cap H}^H(M)$$

for any $(\mathbf{g}, H)$ module V and $(\mathbf{g}, K \cap H)$ module M. A $(\mathbf{g}, H)$ module M is just an $(\mathbf{h}, H)$ module together with a map $\mathbf{g} \otimes M \to M$ of (h, H) modules. The projection formula gives a map $\mathbf{g} \otimes L_K^H(M) = L_K^H(\mathbf{g} \otimes M) \to L_K^H(M)$. Thus $L_K^H(M)$ is a $(\mathbf{g}, H)$ module, when M is a $(\mathbf{g}, K)$ module.

$\mathcal{L}_K^H(\mathcal{M})$ comes naturally equipped with a $\mathcal{D}_\lambda$-module structure. But we can view it in the same functorial way as above, because the tdo $\mathcal{D}_\lambda$ has a G-action. If $\mathcal{V}$ is a $(\mathcal{D}_\lambda, H)$ module on X and $\mathcal{M}$ is a $(\mathcal{D}_\lambda, K \cap H)$ module on X, we have the projection formula:

$$\mathcal{L}_{K\cap H}^H(\mathcal{V} \otimes \mathcal{M}) = \mathcal{V} \otimes \mathcal{L}_{K\cap H}^H(\mathcal{M}) .$$

In particular if we take $\mathbb{1} = \mathcal{O}$ and if $\mathcal{M}$ is a $(\mathcal{D}_\lambda, K)$ module on X, we obtain a map

$$\mathcal{D}_\lambda \otimes \mathcal{L}_K^H(\mathcal{M}) \longrightarrow \mathcal{L}_K^H(\mathcal{M})$$

which give the same structure of $\mathcal{D}_\lambda$-module on $\mathcal{L}_K^H(\mathcal{M})$ as the one given by the direct definition of $\mathcal{L}_K^H$.

The forgetful functor F obviously commutes with $\Gamma(X, .)$. By parallelism, it is easy to see that the following diagram commutes:

$$\begin{array}{ccc} (\mathcal{D}_\lambda, K) - \text{mod} & \overset{\Gamma(X,.)}{\longrightarrow} & (\mathbf{g}, K) - \text{mod} \\ L_K^H \downarrow & & \downarrow L_K^H \\ (\mathcal{D}_\lambda, H) - \text{mod} & \overset{\Gamma(X,.)}{\longrightarrow} & (\mathbf{g}, H) - \text{mod} \end{array}$$

This proves that L commutes with $\Gamma(X, .)$.

By adjointness, the same is true for L and Δ. Indeed for any $(\mathbf{g}, K)$-module M and any $(\mathcal{D}_\lambda, H)$-module $\mathcal{N}$, we have

$$\begin{aligned} \mathrm{Hom}_{(\mathcal{D}_\lambda, H)}(\Delta_\lambda L_K^H(M), \mathcal{N}) &= \mathrm{Hom}_{(\mathbf{g}, H)}(L_K^H(M), \Gamma(X, \mathcal{N})) \\ &= \mathrm{Hom}_{(\mathbf{g}, K\cap H)}(M, F\Gamma(X, \mathcal{N})) \\ &= \mathrm{Hom}_{(\mathbf{g}, K\cap H)}(M, \Gamma(X, F\mathcal{N})) \end{aligned}$$

$$\begin{aligned} &= \mathrm{Hom}_{(\mathcal{D}_\lambda, K\cap H)}(\Delta_\lambda(M), F\mathcal{N}) \\ &= \mathrm{Hom}_{(\mathcal{D}_\lambda, K\cap H)}(\Delta_\lambda(M), F\mathcal{N}) \\ &= \mathrm{Hom}_{(\mathcal{D}_\lambda, H)}(\mathcal{L}_K^H \Delta_\lambda(M), \mathcal{N}) \, . \ \square \end{aligned}$$

II.5 H-spherical $(\mathcal{D}, K)$-modules.

We continue with the complex connected reductive linear algebraic group G. Suppose σ and θ are two commuting involutions of G with respective to fixed point sets H and K. Let P be a parabolic subgroup of G associated with H by an Iwasawa decomposition, and let X be the flag space of type P. The subgroup P is the complexification of the subgroup denoted P^r in §3. Put $X^0 = H \cdot P$; it is the unique open H-orbit in X. Let $\mathcal{D}_\lambda$ be a sheaf of twisted differential operators on $X, \lambda \in t_p^*$, and consider a $(\mathcal{D}_\lambda, K)$-module $\mathcal{M}$. Recall the functor L from the previous section; we will only use $\mathcal{L}_K^H = L_{K\cap H}^H \circ F_K^{K\cap H}$. Since $H/K \cap H$ is an affine variety, the action morphism $a : H \times_{K\cap H} X \to X$ is affine. If $\mathcal{L}_K^H \mathcal{M}$ would be located in several degrees, we would consider only its zero component. Because of Theorems 1.6.3 and II.4.4, we will assume in this section that $\lambda \in \mathbf{t}_p^*$ is dominant.

5.1 Definition: *$\mathcal{M}$ is H-spherical if* $\Gamma(X, \mathcal{L}_K^H \mathcal{M})^H \neq 0$.

Since H acts semisimply on $\Gamma(X, \mathcal{L}_K^H \mathcal{M})$, it is equivalent to say that $\mathcal{M}$ is H-spherical if $\mathrm{Hom}_H(\Gamma(X, \mathcal{L}_K^H \mathcal{M}), \mathbf{C}) \neq 0$. But $\mathcal{L}_K^H$ commutes with $\Gamma(X, .)$, so this is still equivalent to:

$$\mathrm{Hom}_H(L_K^H \Gamma(X, \mathcal{M}), \mathbf{C}) = \mathrm{Hom}_{(\mathbf{h}, K\cap H)}(\Gamma(X, \mathcal{M}), \mathbf{C}) \neq 0 \, .$$

Thus our definition of H spherical module is compatible with the natural one for $(\mathbf{g}, K)$-modules.

For a point x in X, and a subgroup R of G, we denote by R_x the isotropy group of x in R; $R_x = R \cap P_x$. Recall that $\mathcal{M}$ is a $(\mathcal{D}_\lambda, K)$-module.

5.2 Definition: *We say that $\mathcal{M}$ has trivial H-isotropy if for every point $x \in supp\, \mathcal{M}$, we have $\lambda - \rho_p \equiv 0$ on $\mathbf{h}_x$ and $K_x \cap H$ acts trivially on the fiber $j^! \mathcal{M}$ of $\mathcal{M}$ at x, where $j : x \to X$ is the natural inclusion.*

Let $\mathbf{a}_x = \mathbf{t}_{p_x} / \mathbf{h}_x \cap \mathbf{t}_{p_x}$. Then if $\mathcal{M}$ has trivial H-isotropy, $\lambda \in \mathbf{a}_x^*$ at the point x.

5.3 Theorem: *Suppose that $\lambda + \rho_L$ is B-dominant and regular. Let $\mathcal{M}$ be a standard or simple $(\mathcal{D}_\lambda, K)$-module on X corresponding to the data (Y, τ). $\mathcal{M}$ is H-spherical if and only if $Y \cap X^0 \neq \emptyset$ and $\mathcal{M}$ has trivial H-isotropy.*

Proof: As for $(\mathcal{D}, K)$-modules, one can easily see that the open H-orbit X^0 is the only H-orbit which may give rise to a standard $(\mathcal{D}, H)$-module containing the trivial representation of H. By H-equivariance $\dim \mathrm{Hom}_{(\mathcal{O},H)}(\mathcal{O}, \mathcal{L}_K^H(\mathcal{M})) = \dim \mathrm{Hom}_{H_x}(\mathbf{C}, j^\circ \mathcal{L}_K^H(\mathcal{M}))$ where $j : x \hookrightarrow X$ is any point in the open H-orbit X^0. $j^\circ \mathcal{L}_K^H(\mathcal{M})$ is simply the fiber of $\mathcal{L}_K^H(\mathcal{M})$ at x. It is non-zero if and only if $x \in \mathrm{supp}\mathcal{L}_K^H(\mathcal{M}) = H \cdot \mathrm{supp}\mathcal{M} = H \cdot \bar{Y}$. But $\bar{Y} \cap X^0 \neq \emptyset$ is equivalent to $Y \cap X^0 \neq \emptyset$ because X^0 is open. When this condition is satisfied, there may exists a non-trivial map $\mathbf{C} \to j^\circ \mathcal{L}_K^H(\mathcal{M})$ if and only if H_x acts trivially on $j^\circ \mathcal{L}_K^H(\mathcal{M})$. By the definition of $\mathcal{L}_K^H$ this is equivalent to τ being trivial on $(\mathbf{h}_x, K_x \cap H)$. □

Note that in general $j^! \mathcal{L}_K^H \mathcal{M} \neq \mathcal{L}_{K_x}^{H_x} j^! \mathcal{M}$. Let us examine the simplest example.

5.4 Example: $SL_2(\mathbf{R})/\mathbf{R}^*$

$$G = SL_2 \qquad H = \mathrm{diag}\mathbf{C}^* \qquad K = \left\{ \begin{pmatrix} a & b \\ b & a \end{pmatrix} \mid a^2 - b^2 = 1 \right\}$$

$X = \mathbf{P}^1$ with homogeneous coordinates $[z_0, z_1]$. K has three orbits: $Y_1 = \{[1,1]\}$, $Y_{-1} = \{[1,-1]\}$ and the complement Y_0. H has also three orbits: $\{[1,0]\}, \{[0,1]\}$ and the complement X^0. Take $x = [1,2] \in Y_0 \cap X^0, j : x \hookrightarrow X$.

Take $\mathcal{M} = i_* \mathcal{O}_{Y_0}$ for $i : Y_0 \hookrightarrow X$ and $\lambda = \rho$, i.e. $\mathcal{M}$ corresponds to the principal series representation of $SL_2(\mathbf{R})$ containing the trivial representation as a submodule, and every even K-type appears once. Then it is easy to compute that $j^! \mathcal{L}_K^H(\mathcal{M}) = \mathbf{C}^3$. But, $j^! \mathcal{M} = \mathbf{C}$ and $H_x = K_x = \{\pm 1\}$. So $\mathcal{L}_K^H$ does not commute with $j^!$. On the other hand if $\lambda \in \mathbf{t}^*$ is not integral, then $\mathcal{L}_K^H(i_* \mathcal{O}_{Y_0}(\lambda))$ contains only two H-invariant vectors. This agrees with the fact that for all $\lambda \in \mathbf{t}^* = \mathbf{C}$

$$A_{SO(2)}(SL_2(\mathbf{R})/\mathbf{R}^*; \chi_\lambda) = \mathcal{P}_\lambda + \mathcal{P}_\lambda^*$$

where $\mathcal{P}_\lambda$ is a principal series $(\mathbf{sl}_2, \mathbf{C}^*)$-module.

The main result of this section is Theorem 5.5 below which gives a cohomological formula to compute the dimension of the trivial H_x-isotopic component in the fibers of $\mathcal{L}_K^H(i_* \mathcal{O}_Y(\lambda, \tau))$. This dimension can be intepreted as a local multiplicity; the global multiplicity being $\dim \Gamma(X, \mathcal{L}_K^H i_* \mathcal{O}_Y(\lambda, \tau))^H$. Consider the diagram:

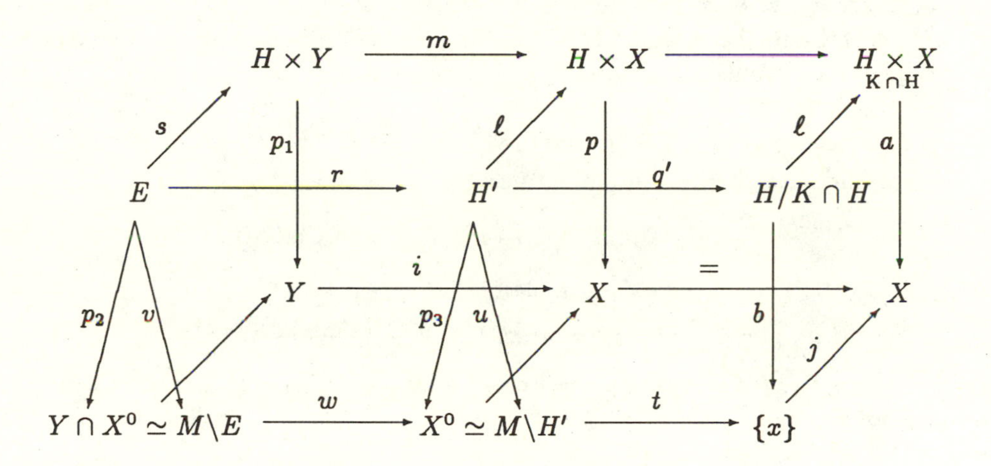

where $x \in Y \cap X^0$ so that $H_x = M$ with $P_x = MAN$. $H' = \{(h, h^{-1} \cdot \mathbf{x}) \in H \times X\}$ and $E = (H \times Y) \cap H' = \{(h, h^{-1} \cdot \mathbf{x}) \in H \times Y\}$. All $k \in K \cap H$ act freely on H' by: $k \cdot (h, H^{-1} \cdot \mathbf{x}) = (hk^{-1}, kh^{-1} \cdot \mathbf{x})$, and $H'/K \cap H$ is the quotient variety. $m \in M$ acts also on H' by $m \cdot (h, h^{-1} \cdot \mathbf{x}) = (mh, h^{-1} \cdot \mathbf{x})$. Obviously these two actions commute. M acts in the same way on E. We start with the $(\mathcal{D}_\lambda, K)$-module $i_* \mathcal{O}_Y(\lambda, \tau)$ and view it as a $(\mathcal{D}_\lambda, K \cap H)$-module. The restriction of $\mathcal{O}_Y(\lambda, \tau)$ to $Y \cap X^0$ gives rise to a $K \cap H$-equivariant local system of rank one that we denote by $\mathbf{C}(\lambda, \tau)$. We will consider the Čech cohomology of this locally constant sheaf.

5.5 Theorem:

$$\mathrm{Hom}_{H_x}(\mathbf{C}, j^\circ \mathcal{L}_K^H i_* \mathcal{O}_Y(\lambda, \tau)) = H^s(Y \cap X^0, \mathbf{C}(\lambda, \tau))^{K \cap H}$$

where $s = \dim Y + \dim M - \dim H \cap K$.

Proof: All the Hom's have to be understood in derived categories. We write d_X for $\dim X$, etc. Combining base change and adjointness we find:

$$\mathrm{Hom}_M(\mathbf{C}, j^\circ \mathcal{L}_K^H i_* \mathcal{O}_Y(\lambda, \tau)[d_X - d_H])$$

$$= \mathrm{Hom}_M(\mathbf{C}, j^! a_* q_+ p^! i_* \mathcal{O}_Y(\lambda, \tau))$$
$$= \mathrm{Hom}_M(\mathbf{C}, b_* \bar{\ell}^! q_+ p^! i_* \mathcal{O}_Y(\lambda, \tau))$$
$$= \mathrm{Hom}_{(\mathcal{D}_{H'/K\cap H,\lambda};M)}(b^* \mathbf{C}, \bar{\ell}^! q_+ p^! i_* \mathcal{O}_Y(\lambda, \tau))$$

q'_+ is an equivalence between the category of $(\mathcal{D}_{H'/K\cap H,\lambda}, M)$-modules on $H'/K\cap H$ and the category of $(D_{H',\lambda}, M \times K\cap H)$-modules on H'; its inverse is q'°. We obviously have $\bar{\ell}^! q_+ = q'_+ \ell^!$. So we obtain:

$$= \mathrm{Hom}_{(\mathcal{D}_{H'/K\cap H,\lambda};M)}(q'^{\circ} b^* \mathbf{C}, \ell^! p^! i_* \mathcal{O}_Y(\tau)) .$$

Applying base change and adjointness again, we find:

$$= \mathrm{Hom}_{(\mathcal{D}_{H'/K\cap H,\lambda};M)}(q'^{\circ} b^* \mathbf{C}, \ell^! m_* p'^! \mathcal{O}_Y(\tau))$$
$$= \mathrm{Hom}_{(\mathcal{D}_{H'/K\cap H,\lambda};M)}(q'^{\circ} b^* \mathbf{C}, r_* s^! p'^! \mathcal{O}_Y(\tau))$$
$$= \mathrm{Hom}_{(\mathcal{D}_{B,\lambda};M\times K\cap H)}(r^* q'^{\circ} b^* \mathbf{C}, (p' \circ s)^! \mathcal{O}_Y(\tau)) .$$

Using the equalities: $p \circ s = c \circ p_2$, $b^{\circ}[d_{K\cap H} - d_H] = b^!$ and $q'^{\circ} b^{\circ} = u^{\circ} t^{\circ}$, we obtain:

$$= \mathrm{Hom}_{(\mathcal{D}_{B,\lambda};M\times K\cap H)}(r^* q'^{\circ} b^{\circ} \mathbf{C}[-d_H + d_{K\cap H}], p_2^! c^! \mathcal{O}_Y(\tau))$$
$$= \mathrm{Hom}_{(\mathcal{D}_{H',\lambda};M\times K\cap H)}(u^{\circ} t^{\circ} \mathbf{C}, r_* p_2^! c^! \mathcal{O}_Y(\tau)[d_{K\cap H} - d_H]) .$$

Using base change again and adjointness again, one gets:

$$= \mathrm{Hom}_{(\mathcal{D}_{X^0,\lambda};K\cap H)}(p_{3!} u^{\circ} t^{\circ} \mathbf{C}, w_* c^! \mathcal{O}_Y(\lambda)[d_{K\cap H} - d_H]) .$$

Now $t^{\circ}\mathbf{C}$ has an action of M, hence $u^! t^{\circ}\mathbf{C} \simeq p_3^! t^{\circ}\mathbf{C}$. Applying $p_{3!}$ and the fact that p_3 is a smooth morphism we see that $p_{3!} u^! t^{\circ}\mathbf{C} \simeq t^{\circ}\mathbf{C}$. It follows:

$$= \mathrm{Hom}_{(\mathcal{D}_{X^0,\lambda},K\cap H)}(t^{\circ}\mathbf{C}, w_* c^! \mathcal{O}_Y(\lambda)[d_H - d_{K\cap H} + d_M])$$
$$= \mathrm{Hom}_{(\mathcal{D}_{Y\cap X^0,\lambda},K\cap H)}(w^* t^{\circ}\mathbf{C}, c^! \mathcal{O}_Y(\lambda)[d_H - d_{K\cap H} + d_M])$$
$$= \mathrm{Hom}_{(\mathcal{D}_{Y\cap X^0,\lambda},K\cap H)}(w^{\circ} t^{\circ}\mathbf{C}[d_{K\cap H} - d_E - d_M], c^{\circ} \mathcal{O}_Y(\lambda)) .$$

Let us recall that there is a shift by $[d_X - d_H]$ on the right-hand term coming from the definition of $\mathcal{L}$. Moreover $d_E = d_Y + d_M$. Putting all the shifts on one side we find:

$$\mathrm{Hom}_{(H)}(\mathbf{C}, j^{\circ} \mathcal{L}_K^H i_* \mathcal{O}_Y(\lambda))$$
$$\mathrm{Hom}_{(\mathcal{D}_{Y\cap X^0,\lambda},K\cap H)}(\mathcal{O}_{Y\cap X^0}, \mathcal{O}_{Y\cap X^0}(\lambda)[d_X - d_H + d_Y + 2d_M - d_{K\cap H}]) .$$

Finally $d_X + d_M = d_H$, and using the Riemann-Hilbert correspondence the above space consists precisely of the $K \cap H$-invariant elements in:

$$\mathrm{H}^s(Y \cap X^0, \mathbf{C}(\lambda))$$

where $s = d_Y + d_M - d_{K\cap H}$ and $\mathbf{C}(\lambda)$ is the local system on $Y \cap X^0$ which corresponds to the $\mathcal{D}$-module $\mathcal{O}_Y(\lambda)$. The dimension of this cohomology space depends on the monodromy of $\mathbf{C}(\lambda)$. □

Note that by an easy homotopy argument, only the action of the group of connected components of $K \cap H$ matters.

Note also that $K \cap H$ is transitive on the connected components of $Y \cap X^0$, at least if $Y \cap X_o \neq \emptyset$, where X^r is the real flag variety of the real form G^r of G, *cf.* [Matsuki, p. 332].

5.6 Examples: Consider again SL_2 as in 5.4. If we take $i : Y_1 = \{[1,1]\} \to X$, then E consists of two points which are permuted by $M = K \cap H = \mathbf{Z}/2$. Thus $\mathrm{Hom}_M(\mathbf{C}, j^! \mathcal{L}_K^H(i_* \mathbf{C}_{Y_1}(\lambda)) = \mathbf{C}$ if $\lambda \in \hat{K}$ is trivial on $H \cap K = \{\pm 1\}$, and $= 0$ otherwise. On the other hand, if we take Y_0 the open K-orbit and $x = [1,2]$, then $Y_0 \cap X^0$ is isomorphic to $C \setminus \{1,\ 0,\ -1\}$. The monodromy given by λ acts around the point 0. H_x and $K \cap H$ are both $\{\pm 1\}$ and they act trivially. Hence $H^0(Y \cap X^0, \mathbf{C}(\lambda)) = \mathbf{C}$ if $\lambda \in \mathbf{Z}$ and is 0 otherwise. Since the Euler characteristic is insensitive to the monodromy we see that $H^1(Y_0 \cap X^0, \mathbf{C}_\lambda) = \mathbf{C}^3$ if $\lambda \in \mathbf{Z}$ and equals $\mathbf{C}^2$ otherwise. This is the multiplicity of the trivial H-type in $\mathcal{L}_K^H(i_* \mathcal{O}_{Y_o})$.

Note that if $K = G$ (compact symmetric space case), or $K = H$ (Riemannian symmetric space case), then the shift s in theorem 5.5 is 0. Moreover in both cases, the only choice for $Y \cap X^0$ is X^0 itself. From this follows easily that the cohomology space in question is at most one dimensional. In other words the corresponding symmetric spaces are multiplicity free in their whole spectrum, compare with theorem 3 of the introduction.

Another interesting example is for the symmetric space $SL_3(\mathbf{R})/S(O_2 \times O_1)$. Then X is the full flag variety of $G = SL_3(\mathbf{C})$, and $K = SO_3(\mathbf{C})$ as only one closed orbit, *cf.* I.7.2. Its intersection with the open orbit of $H = S(GL_2(\mathbf{C}) \times GL_1(\mathbf{C}))$ is isomorphic to $\mathbf{C}^*$ on which $K \cap H$ is transitive. The shift s equals 1, and so does the dimension of the cohomology group when the weight λ is chosen to give rise to an H-spherical module.

Finally, an important example is the group manifold $G/H = (G_1 \times G_1)/diag(G_1)$ for a reductive Lie group G_1. Suppose, we are interested in the representation which occur on the real form $G_{1,o}$ containing a maximal compact subgroup $K_{1,o}$. Then we take $K = K_1 \times K_1$, and X is the product $X_1 \times X_1$ of the full flag variety of G_1 with itself. X^0 is the open Bruhat cell, seen as a G_1-orbit on X. A K-orbit Y is the product of two K_1-orbits in X_1. Given a closed K-orbit Y supporting an irreducible $(\mathcal{D}_\lambda, K)$-module $\mathcal{M}$ with λ dominant in $\mathbf{t}^*$ (here $\rho_\ell = 0$), let M^∞ be the corresponding representation of $G_{1,o}$. Then formula 5.5 must reduce to the multiplicity one of $M^\infty \otimes M^{*,\infty}$ considered as a submodule representation of $G1, o \times G_{1,o}$ in $C^\infty(G_{1,o})$. This multiplicity one is also implied by the Plancherel formula of $G_{1,o}$.

5.7 Remark on the component group of $H \cap K$.

Sometimes, one wants to work with the symmetric space G_o/H_o^0 where H_o^0 is the connected component of the real group $H_o = G_o^\sigma$. The space of functions on G_o/H_o^0 is $\operatorname{ind}_{H_o^0}^{G_o}(\mathbb{1}) = \operatorname{ind}_{H_o}^{G_o}(\operatorname{ind}_{H_o^0}^{H_o}(\mathbb{1}))$, where $\operatorname{ind}_{H_o^0}^{H_o}(\mathbb{1})$ is the regular representation of the group of connected components $H_o^{cc} := H_o/H_o^0$. Since G is connected, this is an abelian 2-group, so $\operatorname{ind}_{H_o^0}^{H_o}(\mathbb{1}) = \oplus_{\varepsilon \in \hat{H}_o^{cc}} \mathbf{C}_\varepsilon$; in particular all multiplicities in this sum are equal to one. So the space $C^\infty(G_o/H_o^0)$ is naturally divided into the various subspaces $C^\infty(G_o/H_o, \varepsilon)$ $:= \operatorname{ind}_{H_o}^{G_o}(\mathbf{C}_\varepsilon)$ where ε runs through the dual group $\hat{H}_o^{cc}$. We say that a representation V of G_o is (H_o, ε)-spherical if $\operatorname{Hom}_{H_o}(V, \mathbf{C}_\varepsilon) \neq 0$. This notion can be also studied algebraically thanks to the following facts.

Let G be a reductive complex linear algebraic group defined over $\mathbf{R}$, possibly disconnected. The datum of a real form G_o of G is equivalent to the datum of a semi-involution c of G, i.e. a conjugate linear involution. Thanks to the existence and uniqueness – up to conjugation – of a compact real form for G, E. Cartan knew that the conjugacy classes of involutions of G are in bijection with the conjugacy classes of semi-involutions of G, *cf.* [Bien1]. The bijection goes as follows: any involution σ stabilizes a compact real form G_u, and therefore it σ commutes with the semi-involution c_u defining G_u. Then the conjugacy class of the product $\sigma \cdot c_u$ is the class of semi-involutions corresponding to σ. Conversely, given a semi-involution c defining a real form G_o, consider an involution θ whose fixed point subgroup in G_o is a maximal compact subgroup K_o, (θ is called a *Cartan involution* of G_o). The complex linear extension of θ to the whole group G is the desired involution. Cartan's bijection gives a one-to-one correspondence between the G-conjugacy classes of symmetric subgroups K of G and the G-conjugacy classes of real forms G_o of G. Moreover if K corresponds to G_o by this bijection, we have:

5.8 Lemma: $K^{cc} \simeq G_o^{cc}$ *where the superscript cc denotes the component group.*

Proof: It is well known that a complex algebraic group is connected (also called irreducible) if and only if it is topologically connected. The compact real form K_o is a homotopy retract of K. Therefore K and K_o have the same component group. But K_o is a maximal compact subgroup of G_o, and because G is algebraic, it is also a homotopy retract of G_o. Thus the component group of K_o and G_o coincides.

Let us apply this result to the group $H = G^\sigma$. The involution θ of G such that $K = G^\theta$, commutes with σ; hence, it defines an involution θ_H of H whose fixed point set is $K \cap H$. The real form $H_o = H \cap G_o$ of H is associated to θ_H by Cartan's bijection applied to H. The lemma implies:

$$(K \cap H)^{cc} \simeq H_o^{cc} .$$

If G is simply connected, then H is connected (a result due to R. Steinberg). Otherwise H is the semidirect product of H° with an abelian 2-group, made of elements of order 2 in a maximal σ-split torus, *i.e.* a torus on which σ acts as $-id$. Following [Matsumoto], we can even write H_o as a semidirect product $H_o^0 \cdot H_o^{cc}$ – (this decomposition is not canonical). Extending trivially the action of $H^c c_o$ to H^0, we can define a new group $H_+ := H^0 \cdot (K \cap H)$.

For $\varepsilon \in \widehat{(K \cap H)^{cc}}$, let $\mathbf{C}_\varepsilon$ be the $(\mathbf{h}, K \cap H)$-module on which $\mathbf{h}$ acts trivially and $K \cap H$ acts by ε. Then we say that a $(\mathbf{g}, K)$-module V is (H, ε)-*spherical* if $\mathrm{Hom}_{(\mathbf{h}, K\cap H)}(V, \mathbf{C}_\varepsilon) \neq 0$. To study this notion with the functor L, we must use $L_K^{H_+}$. We have $\mathrm{Hom}_{(\mathbf{h}, K\cap H)}(V, \mathbf{C}_\varepsilon) = \mathrm{Hom}_{H_+}(\mathbf{C}_\varepsilon, L_K^{H_+}(V))$. Thus this definition corresponds exactly to the one for real groups.

The point is that Theorems 5.3 and 5.5 can be formulated for (H, ε)-spherical representations: it suffices to replace the words "trivial H-isotropy" by "(H, ε)-isotropy", i.e. $\lambda - \rho_p \equiv 0$ on $\mathbf{h}_x$ and $K \cap H_x$ acts on the fibers by the representation

$$K \cap H_x \to (K \cap H)^{cc} \xrightarrow{\varepsilon} \mathbf{C}^\times .$$

In 5.5, one replaces $K \cap H$-invariants by the ε-isotropic subspace of $H^\bullet(Y \cap X^0; \mathbf{C}(\tau))$ for the action of $K \cap H$.

5.9 One more example: Let us remark that if $\mathcal{M} = i_* \mathcal{O}_Y(\tau)$ is a standard H-spherical $(\mathcal{D}_\lambda, K)$-module, $\mathcal{N} = i_{!*}\mathcal{O}_Y(\tau)$ may not be H-spherical.

This happens for example if $G_o/H_o \simeq SL_2(\mathbf{R})$, i.e. $G = SL_2 \times SL_2, H =$ diag$SL_2, K = S0_2 \times S0_2, X = \mathbf{P}^1 \times \mathbf{P}^1$. H has two orbits X^0 and X^1 in X and K has nine orbits. Let Y be a K-orbit consisting of a line minus a point, then $i_*\mathcal{O}_Y$ is H-spherical but $i_{!*}$ is not. (Compare with Theorem 6.8).

II.6 $\mathcal{D}$-modules and hyperfunctions.

The goal of this section is to show that a standard $(\mathcal{D}, K)$-modules on the flag space X can be mapped into the sheaf $\mathcal{B}_{X^r}$ of hyperfunctions on X supported on a real analytic subvariety $X^r = G^r/P^r$ which is the largest flag variety of the real form G^r of G. The map we will obtain generalizes the classical imbedding

$$\mathcal{O}_X \hookrightarrow \mathcal{B}_{X^r} .$$

In the case of K-spherical modules, our map will give a natural imbedding of certain standard $(\mathcal{D}, K)$-modules into the sheaf of hyperfunctions on X^r. From there, the module can be mapped by the usual Poisson transform, into the space of functions on the symmetric space G^r/K^r. This completes the circle of ideas which motivated our definition of spherical $\mathcal{D}$-modules.

First let us cross the bridge going to the analytic set-up. If X^{an} is the complex analytic variety corresponding to X, let $\mathcal{A}$ be the sheaf of holomorphic functions on X^{an}, and set $\mathcal{D}^{an} = \mathcal{A} \otimes_{\mathcal{O}} \mathcal{D}$. Since $\mathcal{A}$ is flat over $\mathcal{O}$, the functor ^{an} : $\mathcal{D}$-mod $\rightarrow \mathcal{D}^{an}$-mod is exact and obviously faithful. Moreover X is a projective variety, so we can apply Serre's *GAGA* principle to deduce that ^{an} is an equivalence of categories, and it commutes with all functors of direct and inverse images.

Next let X^r be a real form of X. This means that we look at X as a scheme defined over $\mathbf{R}$, and we take its $\mathbf{R}$-points X. (In our situation $X = G/P$, X^r is simply the variety of parabolic subgroups of type P in some real form G^r of G). We require in addition that the real dimension of X^r be equal to the complex dimension of X. (Hence we require G^r to contain a parabolic subgroup P^r whose complexification is P). Let us regard X^r as a real analytic subvariety of X. The restriction of the sheaf $\mathcal{A}$ to X^r is the sheaf of real analytic functions on X^r; we write $\mathcal{A}_r := j^*\mathcal{A}$ where $j : X^r \rightarrow X$. Let Γ_{X^r} be the functor of local sections with support in X^r : it maps sheaves on X to sheaves on X with support in X^r. Its derived functors are denoted $\mathcal{H}^i_{X^r}$ and called the functors of local cohomology with support in X^r. Let ω_r

be the orientation sheaf of the normal bundle of X^r in X.

6.1 Definition: The sheaf of hyperfunctions on X supported along X^r is

$$\mathcal{B} = \mathcal{H}^n_{X_r}(\mathcal{A} \otimes \omega_r) .$$

Sato's theorem asserts that $\mathcal{H}^i_{X_r}(\mathcal{A} \otimes \omega^r) = 0$ for $i \neq n$. Another way to think of this sheaf is

$$\mathcal{B} = j_*(j^!\mathcal{A} \otimes \omega_r)[n] .$$

6.2 Remark: The functors $j^!$ and j_* used here are functors between $\mathcal{D}_r$-modules on X_r and $\mathcal{D}^{an}$-modules on X^{an}. $\mathcal{D}_r$ is simply the restriction of $\mathcal{D}^{an}$ to X_r; this is well defined because X^r is a real form of X. In the case at hand, one can give a direct description of $\mathcal{D}_r$ as a quotient of $\mathcal{A}_r \otimes_{\mathbf{C}} U(\mathbf{g})$ by an ideal $J_{r,\lambda}$ determined by the character $\lambda - \rho_p$, as is done in the complex case, *cf.* I.4; here $\mathcal{A}_r$ is the sheaf of real analytic functions on X^r. In this way one can develop a theory of real $\mathcal{D}$-modules; in particular the shift used in the definition of $j^!$ is $[\dim X_r - \dim_R(X)] = [-n]$

Let $f : Y \to X$ be a map between two topological manifolds; set $rdf = \dim Y - \dim X$: the relative dimension of f. We shall denote by $\omega_{Y|X} = \omega_Y / f^{-1}\omega_X$ the orientation sheaf of Y relatively to X. This is the sheaf of sections of the line bundle on Y whose transition functions are given by the sign of the Jacobian determinant for the transition functions of the vector bundle $\mathrm{Ker}(df : TY \to TX)$ if f is a submersion, and $N_X(Y) =$ the normal bundle of Y in X if f is an immersion. For a general map, use the factorization into a cofibration followed by a fibration. Recall that the functors f^* inverse image, $f^!$ inverse image with proper support are defined between the derived categories of complexes of $\mathbf{C}$-sheaves on X and Y, having constructible cohomology. The next result can be seen as a generalization of the Thom isomorphism.

6.3 Property: *There is a canonical morphism of functors:*

$$f^* \longrightarrow f^! \otimes \omega_{Y|X}[-rdf]$$

which is an isomorphism when f is smooth.

Proof: We will only need this property for closed imbeddings. To do a little better, we prove it for smooth maps and closed imbeddings. The general case follows by showing that the above morphism is independent of the factorization of f.

Let $\mathcal{S}$ be a $\mathbf{C}$-sheaf on X and assume f is smooth. Then f behaves locally like the projection of a vector bundle $p : E \to X$. The Thom ismorphism

$$H^i_{cv}(E, p^*\mathcal{S}) \simeq H^{i-rdf}(X, \mathcal{S} \otimes p_! \omega_{E|X})$$

is given by integration along the fibers, and its inverse is the multiplication with the Thom class of E, *cf.* [Bott and Tu p. 88]. The suffice cv denotes cohomology with compact support in the vertical directions. In the derived category we have:

$$f_! f^* \mathcal{S} \simeq f_!(f^! \mathcal{S} \otimes \omega_{E|X})[-rdf] \simeq \mathcal{S} \otimes f_! \omega_{E|X}[-rd\, f]$$

and there is a unique isomorphism $\psi : f^*\mathcal{S} \to f^!\mathcal{S} \otimes \omega_{E|X}[-rd\, f]$ such that $f_!(\psi)$ is the identity on $\mathcal{S} \otimes f_* \omega_{E|X}[-rdf]$.

If f is a closed imbedding, then $f^*\mathcal{S}$ is quasi-isomorphic to the cohomology of $\mathcal{S}$ restricted to Y, while $f^!\mathcal{S}$ is quasi-isomorphic to the local cohomology of $\mathcal{S}$ along Y. Let us first take $\mathcal{S} = \mathbf{C}$. Then

This is seen by working in a slice of X transversal to Y and by using the fact that, for a point $y \in X$,

$$H^i_y(X, \mathbf{C}) = H^i(B_y, \partial B_y; \mathbf{C}) = H^i_c(T_y X; \mathbf{C})$$

where B_y is a small ball around y. The Thom isomorphism for the vector bundle $p : N_X Y \to Y$ yields

$$H^i(Y, \omega_{Y|X}) \simeq H^{i-rdf}_{cv}(N_X Y, \mathbf{C}) .$$

Since tensorization by $\omega_{Y|X}$ is clearly involutive, we get:

$$f^*\mathbf{C} \xrightarrow{\sim} f^!\mathbf{C} \otimes \omega_{Y|X}[-rd\, f] .$$

Now for an arbitrary sheaf $\mathcal{S}$, we use Godement's resolution of $\mathbf{C}$ by flasque sheaves: $0 \to \mathbf{C} \to \mathcal{G}_0 \to \mathcal{G}_1 \to \cdots$. Since we work over a field, the $\mathcal{G}_i$ are even injective. $0 \to \mathcal{S} \to \mathcal{S} \otimes \mathcal{G}_0 \to \mathcal{S} \otimes \mathcal{G}_1 \to \cdots$ is a flasque resolution of $\mathcal{S}$. Let Γ_Y be the functor of local sections with support in Y, then there is a natural map: $\mathcal{S} \otimes \Gamma_Y \mathcal{G}_i \to \Gamma_Y(\mathcal{S} \otimes \mathcal{G}_i)$. The complex $\mathcal{S} \otimes \Gamma_Y(\mathcal{G})$ is quasi-isomorphic to $\mathcal{S} \otimes \Gamma_Y(\mathbf{C})$, so we get a natural map.

$$f^*\mathcal{S} \otimes f^!\mathbf{C} \to f^!\mathcal{S}.$$

Using the result established for $\mathbf{C}$, we obtain the desired morphism.

$$f^*\mathcal{S} = f^*\mathcal{S} \otimes f^*\mathbf{C} \xrightarrow{\sim} f^*\mathcal{S} \otimes f^!\mathbf{C} \otimes \omega_{Y|X}[-rd\, f] \longrightarrow f^!\mathcal{S} \otimes \omega_{Y|X}[-rd\, f]\ . \square$$

6.4 Proposition: *Let $i : Y \to X, j : Z \to X$ be two maps between topological manifolds. Let $\mathcal{S}$ be a complex of* $\mathbf{C}$*-sheaves an X having constructible cohomology. Then the above morphism yields a canonical map of sheaves:*

$$i_! i^! \mathcal{S} \longrightarrow j_*(j^! \mathcal{S} \bigotimes \omega_{Z|X})[-rd\, j]\ .$$

Proof: By adjointness we have a map $i_! i^! \mathcal{S} \to \mathcal{S}$. Apply j^* to get $j^* i_! \mathcal{S} \to j^*\mathcal{S}$. On the other hand, the previous property gives as a map $j^*\mathcal{S} \to j^!\mathcal{S} \bigotimes \omega_{Z|X}[-rd\, f]$. By composition we get an element of $\operatorname{Hom}(j^* i_! i^! \mathcal{S}, j^! \mathcal{S} \otimes \omega_{Y|X}[-rd\, f])$. Using adjointness of j^* and j_*, we obtain the desired map. $\square$

The Riemann-Hilbert correspondence translates these results into identical statements for $\mathcal{D}$-modules and hyperfunctions on X.

Now we return to our homogeneous situation. Let σ be an involution of G with fixed point set H, and let G^r be a real form of G which admits σ as a Cartan involution. Let K be another subgroup of G defined by an involution θ commuting with σ. Let $K^r = K \cap G^r$ and $H^r = H \cap G^r$. Thus H^r is a compact Lie group acting transitively on the real flag space X^r.

6.5 Definition: Let $K^{r\prime}$ be the normalizer in K of G^r. Then $K^{r\prime}$ is a finite extension of K^r, and $K^{r\prime} = K^r$ if K^r is compact.

The group $K^{r,\prime}$ acts on X^r. Given a K-orbit Y in X, we intersect it with the real subvariety X^r to obtain Y^r; this is a union of K_1-orbits. Conversely, given a $K^{r\prime}$-orbit Y^r in X^r (or even a K^r-orbit), the action of the complex group K on it gives rise to a single K-orbit Y.

6.6 Example: Take $G = SL_2(\mathbf{C}), H = S0_2(\mathbf{C}), K = \operatorname{diag} \mathbf{C}^*$. Then $X = \mathbf{P}^1(\mathbf{C})$ and $X^r = \mathbf{P}^1(\mathbf{R})$. We have $G^r = SL_2(\mathbf{R})$. $H^r = S0_2, K^r = \operatorname{diag} \mathbf{R}^*$ and $K^{r\prime} \cong \mathbf{R}^* \cup i\mathbf{R}^*$. In homogeneous coordinates, $z \in K$ (resp. $K^{r\prime}$) acts on X (resp. X^r) by $[x_0, x_1] \to [zx_0, z^{-1}x_1]$ for $x_i \in \mathbf{C}$ (resp. $\mathbf{R}$). Therefore K has three orbits: the points $[1,0], [0,1]$, and their complement $\mathbf{C}_0$. Similarly $K^{r\prime}$ has three orbits: the points $[1,0], [0,1]$, and their complement $\mathbf{R}^*$ which is disconnected. Observe that $\mathbf{R}^*$ splits into two orbits for the action of K^r.

Since the inclusion $j : X^r \to X$ is $K^{r\prime}$-equivariant, the group $K^{r\prime}$ acts on $\mathcal{B}$. Consider the subsheaf $\mathcal{B}_{K^{r\prime}}$ on which $K^{r\prime}$ acts algebraically, i.e. the local sections of $\mathcal{B}$ which transforms under $K^{r\prime}$ according to the restriction of an algebraic representation of K. It is proved in [Kashiwara 2] that the sections of such a sheaf $\mathcal{B}_{K^{r\prime}}$ are automatically distributions. We can twist the sheaf $\mathcal{B}$ in the same way as we twist $\mathcal{A}$. Let $\lambda \in \mathbf{t}_p^*$ be integral for simplicity. Put $\mathcal{B}_{K^r}(\lambda) =$ subsheaf of $\mathcal{B}(\lambda) = j_* j^!(\mathcal{A}(\lambda) \otimes \omega_r)[n]$ whose global sections are K^r-algebraic. Let Y be a K-orbit in X such that $Y_r = Y \cap X_r \neq \emptyset$, and the inclusion $i : Y \to X$ is affine, (recall that this latter condition is automatic if Y is closed). Let $\lambda \in \mathbf{t}_p^*$ be integral, and consider the standard $(\mathcal{D}_\lambda, K)$-module $\mathcal{M}(Y, \lambda) = i_! i^! \mathcal{A}(\lambda)$.

6.7 Theorem: *There is natural morphism: $\mathcal{M}(Y, \lambda) \to \mathcal{B}_{K^{r\prime}}(\lambda)$which respects the action of $(\mathcal{D}_r, K^{r\prime})$, and is injective if Y is closed.*

Proof: This follows from 6.3 and 6.4. Indeed the global sections of $\mathcal{M}(Y, \lambda)$ are K-algebraic, hence they are mapped to $K^{r\prime}$-algebraic elements. $\mathrm{rd}\,(f) = \dim_r X_r - \dim_r X = -n$. Note that with the shift, both modules live in degree zero. If Y is closed, then $\mathcal{M}(Y, \lambda)$ is irreducible over $(\mathcal{D}_\lambda, K)$. But by K-equivariance the elements of $\mathcal{M}(Y, \lambda)$ are determined by their restriction to Y_r. Hence this map must be injective. □

In fact, if Y is closed, it follows from [Matsuki]'s classification of orbits that Y_r is a single closed orbit, see lemma 7.3. Hence the image of the $(\mathcal{D}_\lambda, K)$-module $\mathcal{M}(Y, \lambda)$ consists precisely of the hyperfunctions in $\mathcal{B}_{K^{r\prime}}(\lambda)$ which are supported in Y_r. Using this result, we can refine the classification of H-spherical representations. Assume $\lambda \in \mathbf{t}_p^*$ is dominant integral.

6.8 Theorem: *Let $\mathcal{M} = i_! \mathcal{O}_Y(\tau)$ be a $(\mathcal{D}_\lambda, K)$-module with Y a closed K-orbit such that $Y \cap X^r \neq \emptyset$. Suppose (λ, τ) is trivial on $(\mathbf{h}_x, K \cap H_x)$ and $\lambda + \rho_l$ is B-dominant regular. Then $\mathcal{M}$ is H-spherical.*

Proof: Thanks to theorem 6.7, there is a non-trivial map $f : i_! \mathcal{O}_Y(\lambda) \to \mathcal{B}_{K^{r\prime}}(\lambda)$. Moreover $i_{!*} \mathcal{O}_Y(\lambda)$ is the unique irreducible subquotient of $i_! \mathcal{O}_Y(\lambda)$ whose restriction to Y is non-zero. Hence by the flabbiness of $\mathcal{B}$, $f(i_! \mathcal{O}_Y(\lambda))$ is a submodule of $\mathcal{B}_{K^{r\prime}}(\lambda)$. Now using Helgason's and Flensted-Jensen's isomorphisms, the $(\mathbf{g}, K)$-module M corresponding to $i_! \mathcal{O}_Y(\lambda)$ appears as a submodule of $A_{K_o}(G_o/H_o)$. Thus $\mathrm{Hom}_{(\mathbf{h}, K \cap H)}(M, \mathbf{1}) \neq 0$, or equivalently

$\mathrm{Hom}_H(\mathbf{L}^H_K(M), \mathbb{1}) \neq 0$. This implies:

$$\mathrm{Hom}_H(\mathbb{1}, \mathbf{L}^H_K(M)) = \mathrm{Hom}_{(\mathcal{O},H)}(\mathcal{O}, \mathcal{L}^H_K\mathcal{M}) \neq 0. \quad \square$$

We shall see in IV.6 that in general there may be injective morphisms different than the 'diagonal' $f : i_!\mathcal{O}_Y(\lambda) \to \mathcal{B}_{K^{r\prime}}(\lambda)$ introduced here. They are obtained by composing f with the projection of $\mathcal{B}_{K^{r\prime}}(\lambda)$ onto one of its direct summands.

II.7 Closed orbits and discrete series.

When K_o is a maximal compact subgroup of G_o, the closed K-orbits on the full flag variety of G support part of the fundamental series of representations in $C^\infty(G_o)$, *i.e.* the series of representations associated to the most compact Cartan subgroups. When rank $G =$ rank K, this fundamental series is equal to the part supported by the closed K-orbits, and it fills in the eigenspaces of the discrete spectrum of Z: the center of the enveloping algebra, acting on $L^2(G_o)$; hence its is called the *discrete series* of G_o. Flensted-Jensen, Oshima and Matsuki have shown that similar statements hold for symmetric spaces when one consider real flag varieties and this result can be translated to complex flag varieties where one should consider only some particular closed K-orbits, see [Oshima].

In this section we want to recover Oshima-Matsuki's theorem from our formalism in order to explain the choice of these particular K-orbits. We will also prove that the discrete series of a symmetric space has generically multiplicity one.

As before take G with two involutions σ an θ whose fixed point sets are H and K. Let A be a maximally σ-split torus in G, i.e. a torus on which σ acts by $-id$ and which is maximal for this property. Let $L = \mathrm{Cent}(A; G)$, then $L = MA$ where $M = \mathrm{Cent}(A; H)$. A may be much smaller than the center C of L, but the adjoint action of A on $\mathbf{g}$ yields a genuine root system $R(\mathbf{a})$ while in general $R(\mathbf{c})$ is not a root system. Take a set of positive roots $R^+(\mathbf{a})$ and let N be the nilpotent subgroup of G whose Lie algebra is spanned by the root spaces corresponding to $R^+(\mathbf{a})$. Then $P = LN$ is a parabolic subgroup of G and the flag variety X of parabolic subgroups of G conjugate to P is said to be associated to H by the Iwasawa decomposition.

7.1 Definition:

$$\operatorname{rank} G/H = \dim A = \ell .$$

We say that we are in *the equal rank case* if $\operatorname{rank} G/H = \operatorname{rank} K/H \cap K$, i.e. if we can choose $A \subseteq K$.

Let G_o be a real form of G such that $K_o = K \cap G_o$ is compact and put $H_o = H \cap G_o$. For $\lambda \in \mathbf{a}^*$, let $L^2_K(G_o/H_o;\chi)$ be the space of K_o-finite functions on G_o/H_o which are square integrable, and eigenfunctions of $\mathbf{D}(G_o/H_o)$ for the eigencharacter $\chi = \chi_\lambda$. By elliptic regularity applied to the space $K_o\backslash G_o$, these functions are automatically real analytic on G_o, and hence on G_o/H_o. We want to describe this space of functions as a Harish-Chandra module. Let us first recall a result due to [Flensted-Jensen], [Oshima and Matsuki].

7.2 Theorem: *$L^2_K(G_o/H_o;\chi) \neq 0$ for some λ if and only if $\operatorname{rank} G/H = \operatorname{rank} K/H \cap K$. If λ is singular, $L^2_K(G_o/H_o;\chi) = 0$.*

Therefore we will focus on the equal rank case. Let us denote by x the point of X corresponding to P. The open H-orbit in X is $X^0 := H \cdot x$ and $M := H \cap P$ is the stabilizer of x in H. Since $A \subseteq K$, , $Y = K \cdot x$ is a closed K-orbit in X, and $W_G(A) = W_H(A)$. Choose representatives $w_1, \ldots, w_m$ for the cosets in $W(H/K \cap H) := W_H(A)/W_{K\cap H}(A)$. Then all the closed K-orbits in X whose intersection with X° is not empty are of the form $Y_j = Kw_jP$, as $j = 1, \ldots, m$.

Now we present a geometric lemma which shows the consistence of this approach with the one taken in §6; the proof of the first part was kindly communicated to me by T. Matsuki. Let X be any flag variety for G having a real form X^r corresponding to the real group G^r whose maximal compact subgroup is H^r. Note that X^r is a homogeneous space for H^r. As before let X^0 denote the open H-orbit in X.

7.3 Lemma: *Suppose rank G/H = rank $K/H \cap K$. If Y is a closed K-orbit in X, then $Y \cap X^0 \neq \emptyset$ if and only if $Y \cap X^r \neq \emptyset$. Futhermore in this case $Y^r := Y \cap X^r$ is a single K^r-orbit.*

Proof: The implication in one direction (namely the reverse one) is clear, because X^0 is a complex neighborhood of X^r, hence if a closed K-orbit does not intersect X^0, it cannot intersect X^r. To prove the direct assertion, we first observe that the preimage of Y in the full flag variety of G contains a closed orbit, so we may assume that X is the full flag variety. Let B be a

Borel subgroup of G belonging to X^r which is the only closed orbit of G^r in $X = G/B$. It is also useful to identify the orbits of K (resp. H, G^r) in X with the double cosets $K\backslash G/B$, (resp. $H\backslash G/B$, $G^r\backslash G/B$); the closure relations of the double cosets is the same as the one of the orbits. Thus, we have $X^0 = H \cdot B$ and $X^r = G^r \cdot B$.

Consider the decompositions $\mathbf{g} = \mathbf{h}+\mathbf{r} = \mathbf{k}+\mathbf{s}$ of $\mathbf{g}$ in terms of eigenspaces of σ and θ respectively. By the rank hypothesis, we can find a maximal abelian semisimple subspace $\mathbf{a}$ of $\mathbf{k} \cap \mathbf{r}$ which is also maximal abelian in $\mathbf{r}$; $\mathbf{a}$ is isomorphic to the Lie algebra of the maximal σ-split torus A. Choose a Cartan subalgebra $\mathbf{j}_k$ of $\mathbf{k}$ containing $\mathbf{a}$. Let $R(\mathbf{k}, \mathbf{j}_k)^+$ be a positive system of roots of $\mathbf{j}_k$ in $\mathbf{k}$ which is $(-\sigma)-$*compatible*, *i.e.* $\alpha \in R(\mathbf{k}, \mathbf{j}_k)^+$ and $\alpha \neq \sigma\alpha$ implies $-\sigma\alpha \in R(\mathbf{k}, \mathbf{j}_k)^+$. Since Y is a closed double coset in $K\backslash G/B$, there is a unique positive system $R(\mathbf{g}, \mathbf{j}_k)^+$ such that:

$$R(\mathbf{g}, \mathbf{j}_k)^+ \supseteq R(\mathbf{k}, \mathbf{j}_k)^+$$

and that the Borel subgroup B_1 determined by j_k and $R(\mathbf{g}, \mathbf{j}_k)^+$ is conjugate to B by an element x in the double coset Y.

Since $R(\mathbf{k}, \mathbf{j}_k)^+$ is $(-\sigma)$-compatible, we have:

$$(K \cap H) \cdot (K \cap B_1) \text{ is open in H,}$$

and therefore $(K \cap H)xB$ is open in Y. Thus $x \in Y \cap (H \cdot B)$, because $Y \cap X^0 \neq \emptyset$ is by hypothesis open in Y.

On the other hand, we can find a σ-stable Cartan subalgebra $\mathbf{j}$ of $\mathbf{g}$ containing $\mathbf{j}_k$, such that B_1 is the Borel subgroup determined $\mathbf{j}$ and $R(\mathbf{g}, \mathbf{j})^+$. Using the fact that H-orbits and G^r-orbits are associated [Matsuki, p. 333], it follows that:

$$x \in X^0 \iff x \in X^r.$$

Thus we have proved that $x \in Y \cap X^r$.

For the second part, we use the classification of closed orbits. By the rank hypothesis, $A \subseteq K$. [Matsuki]'s parametrization of K^r-orbits on G^r/P^r shows that there are $m = \#W(H/K \cap H)$ closed K^r-orbits on G^r/P^r. We have found that there are also m closed K-orbit which intersect X^0. Thus the assertion follows from the first part of the lemma. □

Thanks to Theorem 5.3, we can describe all the H-spherical standard $(\mathcal{D}, K)$-modules supported closed K-orbits in X. They must be supported on some Y_i and have trivial H-isotropy. It suffices to do the construction for $Y = K \cdot x$. Let $\mathbf{t}_p = \mathbf{p}/\mathbf{p}_1$ be the Cartan factor of $\mathbf{p}$. Then $\mathbf{a} \simeq \mathbf{a}_p = \mathbf{p}/(\mathbf{p}_1 + \mathbf{m})$ and we have a canonical inclusion $\mathbf{a}_p^* \hookrightarrow \mathbf{t}_p^*$. Y is a flag space for K, but since K may not be connected, so is the isotropy group of x in K. However $K_x \cap H$ contains the group $A \cap M$ of elements of order 2 in A.

Let $i : Y \to X$. Let $\tau : K_x \to \mathbf{C}^\times$ be such that $d\tau$ coincides with $\lambda - \rho_p$ on $\mathbf{k}_x \cap \mathbf{t}_p$. Note that since A can be conjugate into K, this force λ to be an integral weight in $\mathbf{t}_p^*$. Let $\mathcal{O}_Y(\tau) = \mathrm{ind}_{K_x}^K(\tau)$ and set $\mathcal{M}(Y,\tau) = i_*\mathcal{O}_Y(\tau)$.

7.4 Proposition:

- *$\mathcal{M}(Y,\tau)$ is an H-spherical $(\mathcal{D}_\lambda, K)$-module if and only if $\lambda \in \mathbf{a}_p^*$ and τ is trivial on $K_x \cap H$.*

- *When j runs through 1 to m and τ through all possibilities, these modules $\mathcal{M}(Y_j,\tau)$ constitute all the irreducible H-spherical $(\mathcal{D}_\lambda, K)$-modules on X, supported on closed K-orbits.*

This is clear by Theorem 5.3 and the fact that $i_! = i_*$. Put $M(Y_j,\tau) = \Gamma(X, \mathcal{M}(Y_j,\tau))$. Now we can formulate Oshima-Matsuki's result.

7.5 Theorem: *In the equal rank case for λ dominant and regular in $\mathbf{a}_p^*$, there is an isomorphism of $(\mathbf{g}, K)$-modules:*

$$\bigoplus_{j=1}^{m} M(Y_j,\tau) \simeq L_K^2(G_o/H_o;\chi)\ .$$

Proof: $\lambda - \rho_p$ is integral in $\mathbf{t}_p^*$ since it is a character of A and T_p surjects on A. Hence there is a line bundle $\mathcal{L}$ and X such that $i^!\mathcal{L} = \mathcal{O}_Y(\tau)$. Let $\mathcal{B}_{K^{\tau\prime}}(\mathcal{L})$ be the sheaf of $K^{\tau\prime}$-algebraic hyperfunctions along X^τ with values in the line bundle $\mathcal{L}$, i.e. $\mathcal{B}_{K^{\tau\prime}}(\mathcal{L}) = \mathcal{H}^n_{X^\tau}(\mathcal{L} \otimes \omega_\tau)$, *cf.* section 6. Let $Y_j^d = Y \cap X^\tau$ and put $\mathcal{B}_{K^{\tau\prime}}(Y_j,\tau)$ be the space of hyperfunctions in $\mathcal{B}_{K^{1\prime}}(\mathcal{L})$ supported in Y_j^d. By Theorem 6.8, there is a bijective morphism of $(\mathcal{D}_\lambda, K^{\tau\prime})$-modules:

$$\mathcal{M}(Y_j,\tau) \longrightarrow \mathcal{B}_{K^{\tau\prime}}(Y_j,\tau)\ .$$

Since λ is dominant, the global section functor is exact and this bijective morphism carries over to global sections. By lemma 7.3, each set Y_j^τ is a single K^τ-orbit, therefore we do not need to use $K^{\tau\prime}$, but only K_o.

Again by the dominance of λ, the Poisson transform is an isomorphism. Combined with Flensted-Jensen's isomorphism this gives an imbedding of $(\mathbf{g}, K)$-modules

$$\bigoplus_{j=1}^{m} M(Y_j, \tau) \hookrightarrow A_K(G/H; \chi)$$

where $A_K(G/H;\chi)$ is the space of real analytic K_o-finite functions on G_o/H_o with eigencharacter χ_λ for $\mathbf{D}(G/H)$.

Now the main result proved in [Oshima-Matsuki] is that a function in $A_K(G/H;\chi_\lambda)$ is square integrable if and only if it is the image of a hyperfunction in $\mathcal{B}_{K_o}(\mathcal{L})$ supported on some closed orbit Y_j^d. Thus the image of the above imbedding is precisely $L^2_K(G_o/H_o;\chi)$. □

7.6 Corollary: *Suppose that $\lambda + \rho_\ell$ is B-dominant. Let V be an irreducible Hilbert space representation of G_o with infinitesimal character $\lambda + \rho_\ell$. Then the multiplicity of V in $L^2(G_o/H_o, \chi_\lambda)$ is at most one.*

Proof: *We have to show that* $\dim Hom_{G_o}(V, L^2(G_o/H_o, \chi_\lambda)) \leq 1$. Let us denote by the same letter V, the Harish-Chandra module of V. If $\mathrm{Hom}(V, L^2(G/H, \chi_\lambda)) \neq 0$, V imbeds into some $\Gamma(X, i_*\mathcal{O}_Y(\lambda)$ by Theorem 7.5, where $\lambda \in \mathbf{a}^*_+$ is P-dominant and i is the inclusion of a closed K-orbit Y into X. If $\lambda + \rho_\ell \in \mathbf{t}^*$ is B-dominant, then the enveloping algebra U of $\mathbf{g}$ generates $\Gamma(X, \mathcal{D}_\lambda)$, *cf.* I.5.6. Therefore $\Gamma(X, i_*\mathcal{O}_Y(\lambda))$ is an irreducible (U, K)-module and for distinct orbits Y's,they are inequivalent because the $(\mathcal{D}_\lambda, K)$-modules $i_*\mathcal{O}_Y(\lambda)$ are inequivalent. Hence $V = \Gamma(X, i_*\mathcal{O}(\lambda))$ and occurs with multiplicity one in $L^2_K(G/H, \chi_\lambda)$. □

This corollary asserts that for almost all infinitesimal characters, in particular the regular ones, the eigenspace representations $L^2(G/H, \chi_\lambda)$ are multiplicity free. Coupled with the fact that when $\lambda + \rho_\ell$ is B-dominant, two such eigenspaces have no equivalent subrepresentations of G, *cf.* IV.3, we obtain a multiplicity one decomposition of the discrete series of G_o/H_o with sufficiently regular infinitesimal character. Even if λ is P-dominant, it may happen when λ is small enough that $\lambda + \rho_\ell$ is not B-dominant. One knows by coherent continuation that the discrete series modules $\Gamma(X, i_*\mathcal{O}_Y(\lambda))$ – which can also be described as some $A_{\mathbf{p}}(\lambda)$, *cf.* III.8.2 – are irreducible (U, K)-modules or zero. as long as $\lambda \in \mathbf{a}^*$ is P-dominant *cf.* [Vogan 1987]. By theorem 7.5, to prove multiplicity one, it suffices to prove that these modules are inequivalent for distinct orbits. In chapter IV, we will prove this property in the remaining singular cases. To finish this section, we give a description of $\Gamma(X, i_*\mathcal{O}(\lambda))$ as K-module.

7.7 Proposition (Blattner Formula): *For $\lambda \in \mathbf{t}_p^*$ dominant and Y a closed K-orbit, there is an isomorphism of virtual K-modules:*

$$\Gamma(X, i_*i^!\mathcal{O}(\lambda)) = \Sigma_{j\geq 0}(-1)^j H^j(Y, \mathcal{O}(S(\mathbf{n}\cap\mathbf{s})\otimes \mathbf{C}_{\lambda-\rho_p}\otimes\wedge^d(\mathbf{n}\cap\mathbf{s})))$$

where $\mathbf{n}$ *is the nilpotent radical of some* $\mathbf{p}\in Y$, $\mathbf{g}=\mathbf{k}+\mathbf{s}$ *and* $d=\dim(\mathbf{n}\cap\mathbf{s})$.

Proof: There is a K-invariant gradation on $i_*\mathcal{O}_Y(\lambda)$ given by the eigenspaces of the Euler operator radial to Y in X, and for which

$$\mathbf{gr}\, i_*\mathcal{O}(\lambda) \simeq S(T_X/T_Y)\otimes\mathcal{O}(\lambda)\otimes\Omega^{-1}_{X/Y}(\lambda)\,.$$

The shift by $\Omega^{-1}_{X/Y}(\lambda)$ is there because to define $i_*\mathcal{O}_Y(\lambda)$, one has to transform $\mathcal{O}_Y(\lambda)$ into the right $\mathcal{D}$-module of top degree differential forms $\Omega_Y(\lambda)$ on Y, then apply i_* and finally multiply back by $\Omega_X^{-1}(\lambda)$. As a K-sheaf on Y, $\Omega^{-1}_{Y/X}(\lambda)\simeq\mathcal{O}(\wedge^d(\mathbf{n}\cap\mathbf{s}))$.

To prove the identity 7.7 it suffice to apply the functor of global sections in the derived category to the above equality of graded modules and to compute the Euler characteristics. $H^j(X, i_*\mathcal{O}_Y(\lambda)) = 0$ for $j>0$ because $\lambda\in\mathbf{a}_+^*$ is P-dominant. So the left hand side is simply $\Gamma(X, i_*\mathcal{O}_Y(\lambda))$. □

From the Borel-Weil-Bott theorem it follows that the representations of K which may occur in $\Gamma(X, i_*i^!\mathcal{O}(\lambda))$ have an extremal weight of the form

$$\nu = \lambda+\rho_n-\rho_c+\Sigma_{\alpha\in R(\mathbf{n}\cap\mathbf{s})}\; n_\alpha\cdot\alpha \qquad \in \mathbf{t}_{\mathbf{k}}^*\,.$$

Here $\mathbf{t}_{\mathbf{k}}^*$ is the dual of a Cartan subalgebra of $\mathbf{k}$ such that $\mathbf{a}^*\subset\mathbf{t}_{\mathbf{k}}^*$; $\rho_n := \rho(\mathbf{n}\cap\mathbf{s}), \rho_c=\rho(\mathbf{n}\cap\mathbf{k})$ denote half the sum of the non-compact roots, resp. compact roots, and $\rho_n+\rho_c=\rho_p$. In particular if $\lambda+\rho_n-\rho_c$ is K-dominant, then $\lambda+\rho_n-\rho_c$ will be the smallest K-type of $\Gamma(X, i_*i^!\mathcal{O}(\lambda))$.

Given an integral weight $\lambda\in\mathbf{a}^*$, it defines a character χ_λ of $\mathbf{D}(G/H)$, there is a $(\mathbf{g}, K)$ isomorphism

$$L^2_K(G_o/H_o;\chi_\lambda)\simeq\bigoplus_{w\in W(H/K\cap H)}\Gamma(X, i_*^w i^w\mathcal{O}(\lambda))$$

where i^w is the inclusion of the closed K-orbit ${}^wY$ into X. Restricting to K, we have by Blattner's formula:

$$\Gamma(X, i_*^w i^{w!}\mathcal{O}(\lambda)) = \Sigma_{j\geq 0}(-1)^j H^j({}^wY, \mathcal{O}_Y(S({}^w\mathbf{n}\cap\mathbf{s})\otimes \mathbf{C}_{w(\lambda-\rho_p)}\otimes \bigwedge^{top}({}^w\mathbf{n}\cap\mathbf{s}))\ .$$

Thus the set of possible K-types is $w(\lambda+\rho_n-\rho_c+\Sigma_{\alpha\in R({}^w\mathbf{n}\cap\mathbf{s})}\ n_\alpha\cdot\alpha)$, where $w\in W_H(\mathbf{a})$ represents a coset in $W(H/K\cap H)$, and w makes $\lambda+\rho_c-\rho_n$ K-dominant. Since ${}^w\mathbf{p}$ is not conjugate to $\mathbf{p}$ by K, those sets of K-types are different. Taking some large K-dominant K-types E_μ of highest weight $\mu=\lambda+\rho_n-\rho_c+\Sigma_{\alpha\in R(\mathbf{n}\cap\mathbf{s})}n_\alpha\,\alpha$ which occurs in $\Gamma(X, i_*i^!\mathcal{O}_X(\lambda))$, we have:

$$\begin{aligned} H^j(Y,\mathcal{O}_Y(\mu)) &= E_\mu \qquad j=0 \\ &= 0 \qquad \text{otherwise} \end{aligned}$$

Since μ and $w\mu$ are not K-conjugate, E_μ and $E_{w\mu}$ are inequivalent. Hence if we knew that this was the only occurence of these K-types, the $(\mathbf{g},K)$-modules $\Gamma(X, i_*i^!\mathcal{O}_X(\lambda)$ and $\Gamma(X, i_*^w i^{w!}\mathcal{O}_X(\lambda))$ would be inequivalent. Recurrently, one could prove that all the constituents of $L^2_K(G_o/H_o;X_\lambda)$ will be inequivalent. Unfortunately Blattner's formula is only an equality of characters, and it may contain many cancellations. In general it is difficult to say if a possible K-type occurs or not in $\Gamma(X, i_*i^!\mathcal{O}_X(\lambda))$. Therefore to prove that $L^2_K(G_o/H_o;X_\lambda)$ has multiplicity one, we will resort to the more sophisticated techniques of chapters III and IV.

II.8 An algebraic Poisson transform.

The purpose of this section is to describe a functor which transforms $(\mathcal{D}_X, K)$-modules on X into $(\mathcal{D}_{G/K}, K)$-modules on G/K where $\mathcal{D}_{G/K}$ is the sheaf of differential operators on G/K. Since G/K is an affine variety, the global sections of $\mathcal{D}_{G/K}$ are generated by the regular functions on G/K and by the enveloping algebra U of $\mathbf{g}$. This functor is analogous to the Poisson transform in its effect on modules. Its definition could be formulated in terms of real flag varieties and real symmetric spaces, and the definitions are related by a restriction morphism, as in section 6. As for the real case, the inverse functor should be a kind of boundary value map using the so-called 'wonderful' compactification of G/K constructed by Oshima, and [De Concini – Procesi]; see [Springer] for an algebraic description of Oshima's compactification.

First we review the Poisson transform. For simplicity, we suppose that $H_o = K_o$ is a maximal compact subgroup of G_o. Consider a real form G_o of G, and a spherical principal series representation of G_o induced from a minimal parabolic subgroup P_o

$$\mathrm{Ind}_{P_o}^{G_o}(\lambda) := \{f : G_o \to \mathbf{C} \mid f(gp) = e^{\lambda-\rho_P}(p)f(g) \text{for } g \in G_o\,, p \in P_o\,, f \in C^\infty(G_o)$$

where $\lambda \in \mathbf{a}^*$ and A_o is a maximal $\mathbf{R}$-split torus of P_o. G_o acts on $\mathrm{Ind}_P^G(\lambda)$ by left translation, denoted ℓ. Let K_o be a maximal compact subgroup of G_o. Then we can decompose $G_o = K_o \cdot A_o \cdot N_o$ and $P_o = M_o \cdot A_o \cdot N_o$. This representation contains exactly one K_o-invariant vector, say v_λ, because by Fröbenius reciprocity, we have:

$$\mathrm{Hom}_{K_o}(\mathbb{1}, \mathrm{Ind}_{P_o}^{G_o}(\lambda)) = \mathrm{Hom}_{K_o}(\mathbb{1}, C^\infty(K_o/M_r)) = \mathrm{Hom}_{M_r}(\mathbf{1},\mathbf{1}) = \mathbf{C}\ .$$

v_λ can be explicitly written:

$$v_\lambda : G_o = K_oA_oN_o \to \mathbf{C} : \mathrm{k\ a\ n} \mapsto e^{\lambda-\rho_P}(a)\ .$$

There is a G_o-invariant pairing between $\mathrm{Ind}_{P_o}^{G_o}(\lambda)$ and its contragredient $\mathrm{Ind}_P^G(-\lambda)$ given by integration over K_o.

$$\langle u, v\rangle = \int_K u(k)v(k)dk \text{ for } u \in \mathrm{Ind}_{P_o}^{G_o}(\lambda)\ ,\ v \in \mathrm{Ind}_{P_o}^{G_o}(-\lambda)\ .$$

Hence every vector $v \in \mathrm{Ind}_P^G(-\lambda)$ defines a G_o-equivariant map.

$$S_v : Ind_{P_o}^{G_o}(\lambda) \to C^\infty(G_o)$$
$$S_v(u)(g) = \langle \ell(g^{-1})u, v\rangle \quad g \in G_o u \in \mathrm{Ind}_{P_o}^{G_o}(\lambda)\ .$$

Since v_λ is K-equivariant and equal to 1 in K_o, we obtain an intertwining operator $S := S_{v_{-\lambda}}$

$$S : \mathrm{Ind}_{P_o}^{G_o}(\lambda) \to C^\infty(G_o/K_o)$$
$$S(u)(g) = \langle \ell(g^{-1}))u, v_{-\lambda}\rangle = \textstyle\int_K u(gk)\, dk$$

which is called the Poisson transform. A general property of intertwining operators – discovered by Langlands and Miličič – implies that when λ is dominant and regular (recall that the roots of $\mathbf{n}$ are negative), $\mathrm{Ind}_{P_o}^{G_o}(\lambda)$ has a unique irreducible submodule $J(\lambda)$. In our spherical case, this is true also when λ is singular because the unitary spherical principal series is irreducible [Kostant]. $J(\lambda)$ is the socle of $\mathrm{Ind}_{P_o}^{G_o}(\lambda)$ and it consists of the functions which have the larger growth at infinity. Kostant also showed that $v_\lambda \in J(\lambda)$. Dually $\mathrm{Ind}_{P_o}^{G_o}(-\lambda)$ has a unique irreducible quotient $J'(-\lambda)$. $v_{-\lambda}$ can be viewed as a nonzero vector in $J'(-\lambda)$. Moreover $v_{-\lambda}$ is cyclic for $\mathrm{Ind}_P^G(\lambda)$ in the sense that the span of $G_o \cdot v_{-\lambda}$ is dense in $\mathrm{Ind}_{P_o}^{G_o}(-\lambda)$. It follows that when λ is dominant, the Poisson transform S is injective.

8.1 Observation: A representation of G_o may occur as a subquotient in $C^\infty(G_o/K_o)$ although it has no K_o-invariant vector.

For example, take $G_o = SL_2(\mathbf{R}), K_o = S0(2), P =$ the subgroup of upper triangular matrices in G_o. Then $G_o/P_o = \mathbf{P}^1(\mathbf{R})$ and G_o/K_o is the unit disc $\mathbf{D}$.

$$\text{For} \lambda = \rho_p : \quad Ind_{P_o}^{G_o}(\rho_p) \cong C^\infty(\mathbf{P}^1(\mathbf{R})) \text{ and } v_\lambda(kan) = 1 \, .$$

The G-module structure of $C^\infty(\mathbf{RP}^1)$ is one irreducible submodule $\mathbf{C}$ = the constant functions and two irreducible quotients $\mathcal{D}_2$ and $\bar{\mathcal{D}}_2$ which are square-integrable representations. For $f \in C^\infty(\mathbf{RP}^1)$, $(Sf)(gK) = \int_{S0(2)} f(g^{-1}k)dk$. The image by S of a finite Fourier series $\sum_{n=r}^{s} c_n e^{\mathrm{int}}$ on $\mathbf{RP}^1$ is the harmonic polynominal $\sum_{n=0}^{s} c_n z^n + \sum_{n=1}^{r} c_{-n}\bar{z}^n$ on $D^2 \simeq SL_2(\mathbf{R})/S0(2)$. For K-finite vectors, we have

$$\begin{array}{rclrcl} S(\mathbf{C}) & = & \mathbf{C} & (\mathbf{C})_K & = & \mathbf{C} \\ S(\mathcal{D}_2) & = & z\mathbf{C}[z] & (\mathcal{D}_2)_K & = & 0 \\ S(\bar{\mathcal{D}}_2) & = & \bar{z}\mathbf{C}[\bar{z}] & (\bar{\mathcal{D}}_2)_K & = & 0 \end{array}$$

In fact not only $H_0(\mathbf{k}, \mathcal{D}_2) = 0$, but also $H_i(\mathbf{k}, \mathcal{D}_2) = 0$ for all i, since $H_1(\mathbf{k}, \mathcal{D}_2) = H_0(\mathbf{k}, \bar{\mathcal{D}}_2)^*$. Note that the obvious K-invariant functional on $C^\infty(\mathbf{D})$, namely δ_o = evalution at the origin, is zero on $S(\mathcal{D}_2)$ and $S(\bar{\mathcal{D}}_2)$.

In conclusion, we observe that the K_o-spherical irreducible representations of G_o give only a small part of $C^\infty(G_o/K_o)$, but the Poisson transforms gives a whole eigenspace of the ring of differential operators on G_o/K_o. This is the general situation as was proved by Helgason for K-finite vectors and by [Kashiwara, Kowata, Minemura, Okamoto, Oshima and Tanaka] for representations of G_o. see also [Schlichtkrull].

8.2 Theorem: *$S : \mathcal{B}(G_o/P_o; L_\lambda) \tilde{\rightarrow} \mathcal{A}(G_o/K_o; \chi_\lambda)$ is a G_o-isomorphism when λ is dominant, with inverse a boundary value map β.*

Oshima has constructed a compactification of G_o/K_o for which G_o/P_o appears as a piece of the boundary. We have $\dim G_o/K_o = \dim G_o/P_o + \dim A_o$, thus G_o/P_o is the topological boundary of G_o/K_o if and only if this latter symmetric space has rank one.

We pass to complex algebraic groups G, K, A, P, M. In section 6, we saw how to go from $(\mathcal{D}, K)$-modules on G/P to hyperfunctions on G_o/P_o. Now we want to transform a $(\mathcal{D}, K)$ module on G/P into a $(\mathcal{D}, K)$-module on G/K. By analogy with the real situation, we should work with the diagram:

$$G/P \xleftarrow{\pi} G/M \xrightarrow{\kappa} G/K .$$

κ is affine but not proper: its fibers are isomorphic to the open K-orbit in G/P. Since we want to integrate along the fibers of κ, this may look annoying, but the advantage of an affine map is that the direct image functor is exact.

Let $\mathcal{M}$ be a $(\mathcal{D}_\lambda, K)$-module, then $\kappa_*\pi^!\mathcal{M}$ is a $(\mathcal{D}_{G/K}, K)$-module on G/K with eigencharacter χ_λ with respect to the image of Z in $\mathbf{D}(G/K)$: the algebra of G-invariant differential operators on G/K.

8.3 Definition: $\kappa_*\pi^!$ is the algebraic Poisson transform of G associated to K and P.

[De Concini and Procesi] have studied a compactification of G/K' where K' is the normalizer of K in G, which exhibits G/P as a piece of the boundary. It can be constructed as follows. Consider a K-spherical irreducible representation V of G whose highest weight is regular in $\mathbf{a}^*$. The unique line in V fixed by K determines a point in the associated projective space. The closure of the G-orbit of this point in $\mathbf{P}(V)$ is the desired compactification X of G/K'. It is a smooth projective variety on which G acts with a single open orbit: G/K and a single closed orbit: G/P

$$G/P \xrightarrow{i} X \xleftarrow{j} G/K .$$

An analog for $\mathcal{D}$-modules of the boundary value map is the nearby cycle functor whose monodromy invariant part is: i^*j_* where i and j are the canonical inclusions of G/P and G/K in X. It involves a choice of λ corresponding to the infinitesimal character χ_λ. Since this compactification is constructed using only a maximally θ-split Cartan subgroup A, it is natural to focus on the case $\lambda \in \mathbf{a}^*$. By Theorem 2.2, this is even necessary to deal with K-spherical representations. Then for a dominant choice of λ, I would expect the nearby cycle functor to be the inverse of the algebraic Poisson transform $\kappa_* \pi^!$.

G_o/K_o is a real analytic submanifold of G/K. We can restrict $\kappa_* \pi^! \mathcal{M}$ to G_o/K_o. For K-spherical modules, we obtain a subsheaf of the sheaf of real analytic K-finite functions on G_o/K_o which are eigenfunctions of $\mathrm{D}(G/K)$, so that this algebraic Poisson transform and the real Poisson transform coincide. An application of these considerations would be to define maximal globalizations of $(\mathcal{D}, K)$-modules as W. Schmid does it for Harish-Chandra modules.

Finally, let us observe that if H is different than K, the same considerations allow us to transport $(\mathcal{D}, K)$-modules on G/P to $(\mathcal{D}, K)$ modules on G/H. This is useful because G/H may be the complexification of an indefinite symmetric space, and the real Poisson transform can only be defined by meromorphic continuation when H_o is not compact. Since the fibers of $G/M_H \to G/H$ are isomorphic to the open H-orbit X^0 in G/P, this explains in another way the condition in theorem 5.3 on the support of an H-spherical $(\mathcal{D}, K)$-module.

III Microlocalization and Irreducibility

Consider a flag space X, a sheaf $\mathcal{D}_\lambda$ of twisted differential operators on X and an irreducible $\mathcal{D}_\lambda$-module $\mathcal{M}$. Suppose λ is dominant, then we will show that $M = \Gamma(X, \mathcal{M})$ is an irreducible module over $D_\lambda = \Gamma(X, \mathcal{D}_\lambda)$, or it vanishes. However, $U = U(\mathrm{g})$ may not generate all of D_λ and hence M could be reducible as a U-module.

In this chapter, we obtain a criterion for the irreducibility of M as a U-module (Theorem 7.2). Its proof requires a microlocalization technique which is of independent interest and is explained at the beginning of the chapter. This result is the key ingredient in the proof of the irreducibility of the discrete series modules for symmetric spaces. It is proved for certain closed orbits in [Vogan, 1987] using coherent translation of characters.

III.1 Microlocal Differential Operators

Given a smooth complex algebraic variety X, one defines on the cotangent bundle T^*X the sheaf of $\mathcal{E}_X$ microdifferential operators, see [Kashiwara], or [Schapira]. The sheaf $\mathcal{E}_X$ is in some sense the localization of $\mathcal{D}_X$, just as the sheaf of holomorphic functions is a localization of the sheaf of polynomials. More precisely, a microdifferential operator is invertible wherever its principal symbol does not vanish, [Schapira, I.1.3.4]. The sheaf $\mathcal{E}_X$ is a coherent, noetherian sheaf of rings which has a Zariskian filtration. If $p = (x, 0) \in T^*X$, then

$$gr(\mathcal{E}_{X,p}) \cong \mathcal{O}_{X,x}[\xi_1, \cdots, \xi_n]$$

where $n = \dim X$. If $p = (x, \xi) \in T^*X, \xi \neq 0$, then

$$gr(\mathcal{E}_{X,p}) \cong \mathcal{O}_{P^*X,\tilde{p}}[T^{-1}, T]$$

where P^*X is the projective cotangent bundle of X, and $\tilde{p}$ is the image of p in P^*X. Let $\mathcal{E}_X$-mod denote the category of sheaves of $\mathcal{E}_X$-modules on T^*X which are quasi-coherent over $\mathcal{O}_{T^*X}$. The coherent $\mathcal{E}_X$-modules form an abelian category. Let $\tau : T^*X \to X$ be the projection ; then $\mathcal{E}_X$ is flat over $\tau^{-1}\mathcal{D}_X$. Therefore the microlocalization functor *mic* is exact:

$$\mathrm{mic} : \mathcal{D}_X\text{−mod} \longrightarrow \mathcal{E}_X\text{−mod} : \mathcal{M} \longmapsto \mathcal{E}_X \underset{\tau^{-1}\mathcal{D}_X}{\otimes} \tau^{-1}\mathcal{M} .$$

Moreover the functor *mic* preserves coherency and its image is the subcategory of $\mathcal{E}_X$-modules which are coherent on all of T^*X. Since $\mathcal{E}_X \mid T^*_X X \simeq \mathcal{D}_X$, its inverse functor on this subcategory is simply the restriction to the zero section.

The support of an $\mathcal{E}_X$-module $\mathcal{M}$ is also called its characteristic variety and is denoted by $\operatorname{char}\mathcal{M}$. If $\mathcal{M} = \operatorname{mic}\mathbf{N}, \operatorname{char}\mathcal{M}$ is simply the characteristic variety of $\mathbf{N}$ in the sense of $\mathcal{D}$-modules. The variety $\operatorname{char}\mathcal{M}$ is a closed analytic subset of T^*X, stable for the dilation action of $\mathbf{C}^\times$ on T^*X, hence the adjective conical. Moreover it is involutive for the natural symplectic structure on T^*X, i.e. $(\operatorname{char}\mathcal{M})^\perp \subseteq \operatorname{char}\mathcal{M}$. One can also define the characteristic cycle of $\mathcal{M}$, denoted $[\operatorname{char}\mathcal{M}]$, see [Schapira, p. 79].

An $\mathcal{E}_X$-module $\mathcal{M}$ is called holonomic if it is coherent over $\mathcal{E}_X$ and $\operatorname{char}\mathcal{M}$ is a lagrangean subvariety of T^*X, i.e. $(\operatorname{char}\mathcal{M})^\perp = \operatorname{char}\mathcal{M}$. The subcategory of holonomic $\mathcal{E}_X$-mod is denoted by $\mathcal{E}_X$-mod h. There is a natural duality on $\mathcal{E}_X$-mod h which exchanges left and right $\mathcal{E}_X$-modules:

$$* : \mathcal{M} \to \operatorname{Ext}^n_{\mathcal{E}_X}(\mathcal{M}, \mathcal{E}_X)[n] \,.$$

The property $\operatorname{Ext}^j_{\mathcal{E}_X}(\mathcal{M}, \mathcal{E}_X) = 0$ for $j \neq n = \dim X$ characterizes the holonomic $\mathcal{E}_X$-modules among all coherent ones.

The sheaf $\mathcal{E}_X$ operates on the sheaf $\mathcal{C}_X$ of micro-functions on X. One can also define similar sheaves $\mathcal{A}_M, \mathcal{B}_M, \mathcal{C}_M, \mathcal{D}_M, \mathcal{E}_M$ for a real C^1-manifold M. Then $\mathcal{C}_M$ is a sheaf on T^*M whose corresponding presheaf associates to an open set $V \subset T^*M$ the quotient of $\mathcal{B}_M(M)$ by the space of hyperfunctions on M which are micro-analytic at every point of M, [Kashiwara, p. 14].

As an example, take $X = \mathbf{C}, D_X = \mathbf{C}[z, \partial]$, then $\mathcal{E}_X$ contains the element ∂^{-1} defined away from the zero secton of $T^*\mathbf{C}$ and $(\partial - 1)^{-1} = (\partial^{-1} + \partial^{-2} + \partial^{-3} + \cdots)$. The differential operators of infinite positive order do not belong to $\mathcal{E}_X$.

III.2 Microlocalization of a non-commutative ring.

Let $A = \cup_i A_i$ be a filtered unitary ring such that $A_{-1} = 0$ – and which is complete with respect to the topology induced by this increasing filtration. Suppose that grA is a commutative noetherian ring. Then O. Gabber has defined the localization of A with respect to a multiplicatively closed subset $S \subset grA$. We follow the presentation of [Ginsburg]. Consider the category of homomorphisms $f : A \to B$ such that a

(i) B is a complete $\mathbf{Z}$-filtered unitary ring.

(ii) f preserves the filtration.

(iii) the elements of $(grf)(S)$ are invertible in grB.

There is a universal object $i_S : A \to A_S$ in this category; i.e. for any morphism $f : A \to B$ in this category, there is a commutative diagram.

$$\begin{array}{ccc} A & \to & B \\ \downarrow & & \\ A_S & & \end{array}$$

In particular, A_S is a complete and separate $\mathbf{Z}$-filtered ring. The map i_S : $A \to A_S$ is strongly compatible with the filtrations and

$$gr(A_S) \simeq S^{-1} gr\, A.$$

If $0 \notin S$, then $A_S \neq 0$. The map $\sigma : A \to gr\, A$ is called the principal symbol map. We have defined A_S to have the following property: if the principal symbol of $a \in A$ belongs to S, then $i_S(a)$ is invertible in A_S. Finally, A_S is a flat A-module.

2.1 Remark: For any conic open subset $V \subset$ Spec $gr\, A$, let $S = S(V)$ be the multiplicative set of elements of $gr\, A$ invertible on V, i.e. which do not vanish at any point of V. Then $V \to A_{S(V)}$ is a presheaf defined on the conic open subsets of Spec grA. We denote the corresponding sheaf by $\mathcal{A} = \text{mic}(A)$, and call it the formal microlocalization of A. This terminology agrees with § 1 in the sense that $\mathcal{E}_X$ consists of the elements in mic(D_X) which satisfy a certain convergence property. Also if M is a coherent noetherian A-module, then the sheaf associated to $V \mapsto A_{S(V)} \otimes_A M$ is a coherent sheaf of mic A-modules denoted mic M and is called the formal microlocalization of M. The functor

$$\begin{array}{rlcl} \text{mic}: & A - \text{mod}\, c & \longrightarrow & \mathcal{A} - \text{mod}\, c \\ & M & \longmapsto & \mathcal{A} \otimes_A M \end{array}$$

is exact. The support of mic M is the characteristic variety of M.

2.2. Construction A_S: Let t be a transcendental over A. Consider the graded ring $FA = \Sigma t^i A_i$ where $A = \cup_i A_i$ is the $\mathbf{Z}$-filtration on A. It is a subring of $A[t,t^{-1}]$. When the ground field is $\mathbf{C}$, FA should be regarded as a bundle of rings over $\mathbf{C}^*$ whose fiber at $z \in \mathbf{C}^*$ is $A \simeq FA/(t-z)FA$ and which specializes to $gr\,A \simeq FA/+tFA$ when $z \to 0$. If $b \in A$ is a locally ad-nilpotent element, i.e. for every $a \in A$, there is an $n \in \mathbf{N}$ such that $(adb)^n(a) = 0$ where $adb(a) = ba - ab$, then the localization A_b of A with respect to the multiplicative set $\{b^n \mid n \in \mathbf{N}\}$ is well-defined. For example, the localization of the ring FA at t is isomorphic to $A[t,t^{-1}]$.

Now consider the multiplicatively closed subset S of grA we had at the beginning. Let $FS \subset FA$ be the set of homogenous elements in t^kA_k – for some k – such that their images in $t^kA_k/t^{k-1}A_{k-1} \simeq A_k/A_{k-1} \subset grA$ belong to S. Set $A^{(k)} = FA/t^kFA$ and $S^{(k)} = FS/t^kFA \cap FS$. Using the commutativity of grA, it is easy to check that every $u \in A^{(k)}$ is ad-nilpotent, (in fact $(ada)^{k+1} \mid A^{(k)} \equiv 0$). This guarantees the existence of the non-commutative localization $(S^{(k)})^{-1}A^{(k)}$. The projective system

$$\to FA/t^3FA \longrightarrow FA/t^2FA \longrightarrow FA/tFA \simeq grA$$

combined with the universal property of localization gives rise to the projective system.

$$\to (S^{(3)})^{-1}A^{(3)} \longrightarrow (S^{(2)})^{-1}A^{(2)} \longrightarrow (S^{(1)})^{-1}A^{(1)} \simeq S^{-1}gr\,A\,.$$

Set $B = \varprojlim_i (S^i)^{-1}A^{(i)}$. This ring is the localized version of FA. The equality $A = FA/(t-1)FA$ suggests the definition.

$$\begin{aligned} A_S &= B/(t-1)B \\ &= \varinjlim_{mult.\ by\ t} \varprojlim_i (S^i)^{-1}A^{(i)}\,. \end{aligned}$$

This presentation of A_S is not practical for applications, but the important things to remember are the properties of A_S.

III.3 Microlocalization of $\mathcal{D}_\lambda$

We shall need twisted sheaves of formal micro-differential operators in the same way as $\mathcal{D}_\lambda$ is a twisted version of $\mathcal{D}_X$. Let X be the complex flag space of type P of the group G and let $\lambda + t_{\mathfrak{p}}^*$ be as in chapter I. Set $\tau : T^*X \to X$. We have defined the twisted sheaf of differential operators $\mathcal{D}_\lambda$ on X, cf.

I.4.3. The sheaf $\mathcal{D}_\lambda$ has a natural filtration by degree, and there is a symbol isomorphism $\sigma : gr\,\mathcal{D}_\lambda \xrightarrow{\sim} \tau \cdot \mathcal{O}_{T^*X}$. Applying τ^{-1} we get an injection $\tilde{\sigma}$: $gr(\tau^{-1}\mathcal{D}_\lambda) \to \mathcal{O}_{T^*X}$ which maps $gr(\tau^{-1}\mathcal{D}_\lambda)$ onto the sheaf of germs of regular functions on T^*X which are polynominal in the fiber variables. Fix an affine open subset U in X, then $T^*U \subset T^*X$ can be identified with Spec$(gr\,\mathcal{D}_\lambda(U))$. For a conic open subset V of T^*U such that $\tau V = U$, let $S(V)$ be the multiplicatively closed set of elements of $gr(\tau^{-1}\mathcal{D}_\lambda)(V) = (gr\,\mathcal{D}_\lambda)(U)$ which are invertible on V. Then as in section 2, taking $(gr\,\mathcal{D}_\lambda(U))$ for A, we can define $\mathcal{E}_X(V) := (gr\,\tau^{-1}\mathcal{D}_\lambda)(V)_{S(V)}$. It is easily seen that the open sets V of T^*X which project onto affine open sets of X form a base of the Zariski topology of T^*X. Hence we obtain a presheaf $V \mapsto \mathcal{E}_X(V)$ defined on the set of conic open subsets of T^*X. By definition, the corresponding sheaf $\mathcal{E}_\lambda$ is the formal microlocalization of $\mathcal{D}_\lambda$.

3.1 Definition: *The sheaf $\mathcal{E}_\lambda$ on T^*X constructed above is called a twisted sheaf of formal micro-differential operators - tmo for short.*

The sheaf $\tau^{-1}\mathcal{D}_\lambda$ is subsheaf of $\mathcal{E}_\lambda$. If $D \in \mathcal{D}_\lambda$, $\pi^{-1}D$ is invertible in $\mathcal{E}_\lambda$ wherever its symbol does not vanish. We also have a symbol isomorphism $\sigma : gr\,\mathcal{E}_\lambda \xrightarrow{\sim} \bigoplus_{j\in\mathbf{Z}} \mathcal{O}_{T^*X}(j)$ where $\mathcal{O}_{T^*X}(j)$ denotes the sheaf of germs of homogenous functions of degree j in the fiber variables. Moreover $\mathcal{O}_{T^*X}$ is faithfully flat over $\bigoplus_{j\in\mathbf{Z}} \mathcal{O}_{T^*X}(j)$, [Schapira p. 77], so we will often identify a graded $gr(\mathcal{E}_\lambda)$-module $\mathcal{M}$ with the module $\mathcal{O}_{T^*X} \underset{gr\,\mathcal{E}_\lambda}{\otimes} \mathcal{M}$. From the general properties of microlocalization in §2, we have the following result.

3.2. Property: *The functor* mic : $\mathcal{D}_\lambda - \text{mod}\,c \to \mathcal{E}_\lambda - \text{mod}\,c : \mathcal{M} \to \mathcal{E}_\lambda \otimes_{\pi^{-1}\mathcal{D}_\lambda} \mathcal{M}$ *is exact and faithful. Its image consists of the $\mathcal{E}_\lambda$-modules which are coherent on all of T^*X; on this subcategory, the left inverse functor is the restriction to the zero section of T^*X.*

3.3. Remark: Strictly speaking, microdifferential operators are formal microdifferential operators which satisfy a certain convergence property. If $\sum_{-\infty\leq j\leq m} p_j(x,\xi)$ is the expression in local coordinates of a microdifferential operator D defined over an open set $U \subset T^*X$ and p_j is homogenous of degree j in the variable $\xi = (\xi_1, \dots \xi_n)$. Then $D \in \mathcal{E}_X$, i.e. D is a microdifferential operator over U if for every compact subset $K \subset U$, there exists $\varepsilon > 0$ such that $\sum_{0\leq j} \sup_{(x,\xi)\in K} |p_{-j}(x,\xi)| \frac{\varepsilon^j}{j!}$ is finite. So we want the terms $\sup_{(x,\xi)\in K} |p_{-j}(x,\xi)|$

to be the coefficients of a Taylor series on $\mathbf{C}$ which converges in some disk of radius ε. In this approach, it would be natural to define a sheaf of twisted micro-differential operators as a pair $(\mathcal{E}, i : \pi^{-1}\mathcal{D} \to \mathcal{E})$ locally isomorphic to the standard pair $(\mathcal{E}_X, i_X : \pi^{-1}\mathcal{D}_X \to \mathcal{E}_X)$.

III.4 Microlocalization of U_χ

We adopt again the notation of chapter I. Let Z be the center of the enveloping algebra $U = U(\mathbf{g})$ and pick $\chi \in \operatorname{Max} Z$. Define

$$I_\chi = \{z - \chi(z) \mid z \in Z\}U .$$

Let p be a parabolic subalgebra of $\mathbf{g}$ and define as in I.5:

$$J_P = \operatorname{Ann}_U(U \bigotimes_{U([\mathbf{p},\mathbf{p}])} \mathbf{C}) .$$

I_χ and J_P are both two sided ideals in U. Set

$$U_\chi := U/(I_\chi + J_P) .$$

The natural filtration on U induces a filtration on U_χ. With this filtration, $gr U_\chi$ may be non-reduced, for it may be different from the filtration induced by the order of differential operators on X.

4.1. Lemma: $\mathcal{N}_p := \operatorname{Spec} gr\, U_\chi$ *is the closure of the Richardson nilpotent conjugacy class* C_p *of* $\mathbf{p}$ *in* $\mathbf{g}^*$.

This lemma is proved in [Borho-Brylinski I, p. 456]. They also give a criterion ensuring the coincidence of the two filtrations on U_χ, [p. 459]; see also 6.2 below.

$\mathcal{N}_p$ is a closed cone in $\mathbf{g}^*$ and taking U_χ for A in § 2, we can define the formal microlocalization $\mathcal{U}_\chi$ of U_χ on $\mathcal{N}_p$. We regard it as a sheaf on $\mathbf{g}^*$ supported on $\mathcal{N}_p$. For any conic open subset V of $\mathcal{N}_p$, $\mathcal{U}_\chi(V)$ consists of U_χ together with the inverses of the elements of U_χ whose principal symbol do not vanish at any point of $V \cap \mathcal{N}_p$. In particular, if V contains $0 \in \mathbf{g}^*$, then $\mathcal{N}_p \subset V$ and $\mathcal{U}_\chi(V) = U_\chi$.

III.5 Microlocal study of the moment map.

Let G be a complex connected reductive algebraic group and P a parabolic subgroup of G. Let X be the flag space of G of type P. The cotangent bundle T^*X is isomorphic to the set of pairs $(\mathbf{p}, x)$ such that $\mathbf{p}$ is a parabolic subalgebra of $\mathbf{g}$ conjugate by G to Lie P and $x \in \mathbf{p}^\perp \subset \mathbf{g}^*$. G acts on T^*X by conjugating the pairs $(\mathbf{p}, x)$ and it preserves the canonical symplectic structure of T^*X. This action gives rise to a so-called moment map π which in our case is just the projection

$$\pi : T^*X \longrightarrow \mathbf{g}^* : (\mathbf{p}, x) \longmapsto x.$$

π is G-equivariant with respect to the coadjoint action on $\mathbf{g}^*$. The image of π is $\mathcal{N}_p = G.\mathbf{p}^\perp$ = the closure of the Richardson nilpotent conjugacy class C_p of P in $\mathbf{g}^*$.

Now consider an admissible subgroup K of G, that is a subgroup having only finitely many orbits on the full flag variety of G, and let $\mathbf{k}$ be its Lie algebra. Let $\mathcal{M}$ be a coherent $(\mathcal{D}_\lambda, K)$ module on X for any $\lambda \in \mathbf{t}_p^*$, cf. Chapter 1; and let $M = \Gamma(X, \mathcal{M})$ be the corresponding coherent $(\mathbf{g}, K)$ module. If $\lambda + \rho_\ell$ is dominant and regular, then M generates $\mathcal{M}$ over $\mathcal{D}_\lambda$, cf. I.6. Let Char $\mathcal{M} \subset T^*X$ be the characteristic variety of $\mathcal{M}$. The support of M – denoted Supp M, also called the associated variety of M – is the subvariety of $\mathbf{g}^*$ defined by the annihilator of $gr\, M$, for a good filtration on M. Finally let W be the set of K-orbits on X.

5.1. Lemma:

(a) Char $\mathcal{M} \subseteq \cup_{Y \in W} \overline{T_Y^* X}$

(b) $\pi(\overline{T_Y^* X}) = \overline{K(\mathbf{p}^\perp \cap \mathbf{k}^\perp)}$ *for all* $\mathbf{p} \in Y$

(c) *If* $\lambda + \rho_\ell$ *is dominant, integral and regular, then* $\pi(\text{Char}\, \mathcal{M}) = \text{Supp}\, M$.

Proof: These results are proved for $\lambda = \rho_\ell$ in [Borho-Brylinski III]: see 2.5 for (a), 2.4 for (b), 1.9 for (c). (a) and (b) are independent of λ. The translation functors give equivalence of categories inside the dominant chamber; this implies (c). □

Let $\chi = \chi_\lambda$ be the infinitesimal character defined by $\lambda \in \mathbf{t}_p^*$ and the Harish-Chandra morphism $\psi_p : \mathbf{t}_p^* \to \text{Spec}\, Z$. Consider the algebra $U_\chi =$

$U/I_\lambda + J_{\mathbf{p}}$ recalled in §4. Then M is naturally a (U_χ, K)-module. We also have the sheaf of algebras $\mathcal{U}_\chi = \mathrm{mic}\ U_\chi$ on $\mathcal{N}_p \subset \mathbf{g}^*$.mic M is a sheaf of (U_χ, K)-modules whose support is contained in $\mathcal{N}_p \subset \mathbf{g}^*$. Let π_* and π^* denote the direct and inverse images between $\mathcal{E}_\lambda$-modules on T^*X and $\mathcal{U}_\chi$-modules on $\mathbf{g}^*$.

5.2. Proposition: *The following diagram commutes.*

$$\begin{array}{ccc}
\begin{array}{c}(\mathcal{E}_\lambda, K)-\mathrm{mod}\, c \\ \text{on } T^*X\end{array} & \underset{\pi^*}{\overset{\pi_*}{\rightleftarrows}} & \begin{array}{c}(\mathcal{U}_\chi, K)-\mathrm{mod}\, c \\ \text{on } \mathcal{N}_p\end{array} \\
\mathrm{mic}\uparrow & & \uparrow\mathrm{mic} \\
\begin{array}{c}(\mathcal{D}_\lambda, K)-\mathrm{mod}\, c \\ \text{on } X\end{array} & \underset{\Delta_\lambda}{\overset{\Gamma}{\rightleftarrows}} & \begin{array}{c}(U_\chi, K)-\mathrm{mod}\, c \\ \text{on } \{0\} \in \mathbf{g}\end{array}
\end{array}$$

5.3. Remark: For $\mathcal{M} \in \mathcal{D}_\lambda$-mod, $\Gamma\mathcal{M}$ is first a module over $D_\lambda := \Gamma\mathcal{D}_\lambda$. Since $\mathcal{D}_\lambda$ is a *t.d.o* with a G-action, we have a map $U \to D_\lambda$ which factors through U_χ. Hence by "restriction of scalars" we may view $\Gamma\mathcal{M}$ as a U_χ-module. Note that the commutativity does not say what are the composite functors $\Delta_\lambda.\Gamma$ or $\pi_*\pi^*$.

Proof: If we call $\gamma : X \to \{e\}$ the map of X onto the point e, then Γ is γ_* and Δ_λ is γ^*. The following diagram commutes.

$$\begin{array}{ccc}
T^*X & \xrightarrow{\pi} & \mathcal{N}_p \\
\tau\downarrow & & \downarrow\xi \\
X & \xrightarrow{\gamma} & \{e\}\,.
\end{array}$$

We have the projection formulas for a $(\mathcal{D}_\lambda, K)$-module $\mathcal{F}$ and a $(\mathcal{U}_\chi, K)$-module F:

$$\pi_*(\pi^*\mathcal{U}_\chi \otimes_{\tau^{-1}\mathcal{D}} \tau^{-1}\mathcal{F}) = \mathrm{mic}\,\Gamma(X, \mathcal{F})$$
$$\pi^*(\pi_*\mathcal{E}_\lambda \otimes_{\xi^{-1}\mathcal{U}_\chi} \xi^{-1}F) = \mathrm{mic}\,\Delta_\lambda F\,.$$

The definitions of the functor *mic* imply now the proposition.

III.6 Associated variety of a submodule of $\Gamma\mathcal{M}$

First recall a useful result proved in I.6.6.(1). Let $\mathcal{D}$ be a *t.d.o.* on X.

6.1. Lemma: *Let* $D = \Gamma(X, \mathcal{D})$. *If* Γ: $\mathcal{D}-\mathrm{mod\,c} \to D-\mathrm{mod\,c}$ *is exact, then it sends simple objects to simple ones, or to zero.*

By restriction of scalars, any D_λ module on X is a U_χ-module for $\chi = \chi_\lambda \in \mathrm{Max}\, Z$. We are interested in the following problems: let $\mathcal{M}$ be an irreducible $\mathcal{D}_\lambda$-module. When is $\Gamma\mathcal{M}$ irreducible over U_χ? When is $\Gamma\mathcal{M}$ completely reducible over U_χ? U_χ considered as a subalgebra of D_λ inherits a filtration from the operator filtration of D_λ. This filtration is in general different from the natural filtration of U_χ induced by the filtration of U. However, we have the following result:

6.2 Proposition:
(a) *If the moment map* $\pi : T^*X \to C \subset \mathbf{g}^*$ *is birational and has a normal image, then* $U_{\chi_\lambda} = D_\lambda$ *for every* $\lambda \in t_{\mathbf{p}}^*$; *and the filtrations coincide.*
(b) *If* $\lambda + \rho_\mathfrak{l}$ *is dominant in* $\mathbf{t}^*$, *then* $U_{\chi_\lambda} \xrightarrow{op} D_\lambda$ *is surjective for any flag space* X. *In particular,* $U_{\chi_{\rho_\mathbf{p}}} \xrightarrow{op} D_X$ *is always an isomorphism.*

Proof: Part (a) is due to Beilinson, Bernstein. The result follows from the fact that the map $S(\mathbf{g})/_{grI_X + grJ_P} \xrightarrow{\sim} \Gamma(X, gr\,\mathcal{D}_\lambda)$ is an isomorphism, which is independent of λ.

We proved part (b) in I.5.6. The final statement was first proved in [Borho-Brylinski I, p. 452], thanks to the observation that D_λ is always equal to the ring of $\mathbf{g}$-finite endomorphisms of the generalized Verma module $M_P(\lambda) = U(\mathbf{g}) \otimes_{U(\mathbf{p})} \mathbf{C}_{\lambda - 2\rho_\mathbf{p}}$. □

Note that when $\mathcal{N}_\mathbf{p}$ is not normal, or π not birational, the natural filtration on U_{χ_λ} may be different from the operator filtration on $\Gamma(X, \mathcal{D}_X)$. Proposition 6.2 answers our first problem in most cases: namely, the U_χ-module $\Gamma\mathcal{M}$ is irreducible. However, some examples which escape the hypothesis of 6.2, are discrete series representations on symmetric spaces. So it would be a shame to disregard them.

Applying the technique of microlocalization, we can go one step further than proposition 6.2. Let us consider a closed conic Lagrangean subvariety Λ of $\cup_{Y \in W} \overline{T_Y^* X}$ where W is the set of K-orbits in X. Put $Z = \pi(\Lambda) \subseteq \mathcal{N}_\mathbf{p}$. Recall that the generic point Z° of a scheme Z is any point whose closure is Z. One can think of Z° as being an open subset of Z whose closure is Z. We say that the moment map $\pi : T^*X \to \mathcal{N}_p$ is an isomorphism in a neighborhood of Λ° if there is a neighborhood Λ' of Λ° in T^*X and a neighborhood Z' of Z° in $\mathcal{N}_p$ such that $\pi : \Lambda' \to Z'$ is an isomorphism. Let $\mathcal{M}$ be an $\mathcal{E}_\lambda$ (resp U_χ) -module on T^*X (resp. $\mathcal{N}_p$) with support Λ (resp. Z). We say that $\mathcal{M}$ is irreducible at its generic point if $\mathcal{M} \mid \Lambda^\circ$ (resp. Z°) is irreducible over $\mathcal{E}_\lambda \mid \Lambda^\circ$ (resp $\mathcal{U}_\chi \mid Z^\circ$). It is clear that if $\mathcal{M}$ is irreducible then it is irreducible at its generic point. Now let $\mathcal{M}$ be an irreducible $\mathcal{E}_\lambda$-module whose support is Λ and suppose $\pi_*\mathcal{M} \neq 0$.

6.3 Lemma: *If π is an isomorphism in a neighborhood of $\mathring{\Lambda}$ and if λ is P-dominant, then $\pi_*\mathcal{M}$ is irreducible at its generic point.*

Proof: The first hypothesis means that π is birational and that Z contains an open dense subset of normal points of $\mathcal{N}_p$. We can take $\mathring{Z}$ to be the open subset of Z made of normal points of $\mathcal{N}_p$ above which the fiber of π consists of a single point. Then let $\mathring{\Lambda} = \pi^{-1}(\mathring{Z})$, so that $\pi : \mathring{\Lambda} \to \mathring{Z}$ is a bijective map. Let Λ' and Z' be their neighborhoods in T^*X and $\mathcal{N}_p$ such that $\pi : \Lambda' \to Z'$ is an isomorphism. Then $\mathcal{E}_\lambda(\Lambda') \simeq \mathcal{U}_\chi(Z')$. Moreover since λ is P-dominant, π_* is exact with inverse π^*. Hence $(\pi_*\mathcal{M}) \mid \mathring{Z} \simeq \pi_*(\mathcal{M} \mid \mathring{\Lambda})$ is irreducible over $(\pi_*\mathcal{E}_\lambda) \mid \mathring{Z} \simeq \mathcal{U}_\chi \mid \mathring{Z}$. □

The associated variety $\mathrm{Ass}(M)$ of a Harish-Chandra module M is the zero set of the ideal $\mathrm{Ann}(grM)$ in $S(\mathbf{g})$ for a good filtration on M. When M is a U_χ-module, we have $\mathrm{Ass}(M) = \mathrm{supp}(\mathrm{mic}\, M)$.

6.4 Theorem: *Let $\mathcal{M}$ be an irreducible $\mathcal{D}_\lambda$-module on X with λ being P-dominant in $\mathbf{t}_p^*$. Let $N \to \Gamma\mathcal{M}$ be a U submodule of $\Gamma\mathcal{M}$. Then:*

$$\mathrm{Ass}(N) = \mathrm{Ass}(\Gamma\mathcal{M}) .$$

Proof: $\mathrm{Hom}_{(\mathbf{g},K)}(N, \Gamma\mathcal{M}) = \mathrm{Hom}_{(\mathcal{D},K)}(\Delta N, \mathcal{M})$. Thus we get a nonzero map: $\Delta N \to \mathcal{M}$, which must be surjective since $\mathcal{M}$ is irreducible. By hypothesis Γ is exact, hence $\Gamma\Delta N \to \Gamma\mathcal{M}$ is still surjective and the composite map $N \to \Gamma\Delta N \to \Gamma\mathcal{M}$ is the original inclusion. Looking at associated varieties we get:

$$\mathrm{Ass}(N) \subseteq \mathrm{Ass}(\Gamma\mathcal{M}) \subseteq \mathrm{Ass}(\Gamma\Delta N). \tag{1}$$

Now $\mathrm{Ass}N = \mathrm{supp}(\mathrm{mic}\, N)$ and $\mathrm{Ass}(\Gamma\Delta N) = \mathrm{supp}(\pi_*\pi^*\mathrm{mic}\, N)$. But $\mathrm{supp}(\pi^*\mathrm{mic}\, N) \subseteq \pi^{-1}(\mathrm{supp}\,\mathrm{mic}\, N)$ and $\mathrm{supp}(\pi_*\mathcal{F}) \subseteq \pi(\mathrm{supp}\mathcal{F})$ for any $\mathcal{E}_\lambda$-module for T^*X. Hence $\mathrm{supp}(\pi_*\pi^*\mathrm{mic}\, N) \subseteq \pi\pi^{-1}(\mathrm{supp}\,\mathrm{mic}\, N) = \mathrm{supp}(\mathrm{mic}\, N)$. Comparing this inclusion with (1), we get

$$\mathrm{Ass}(N) = \mathrm{Ass}(\Gamma\mathcal{M}) = \mathrm{Ass}\Gamma\Delta N\,. \ \square$$

III.7 Irreducibility criterion

We can now reap the harvest of microlocalization. Let $\mathcal{M}$ be a simple $(\mathcal{D}_\lambda, K)$-module on X with characteristic variety Λ. Suppose $\Gamma(X,\mathcal{M}) \neq 0$, and λ is P-dominant (so that the functor Γ is exact),

7.1 Proposition: *If π is an isomorphism in a neighborhood of $\mathring{\Lambda}$, then every proper quotient of $\pi_*(\mathrm{mic}\mathcal{M})$ is supported on a subset of $Z = \pi(\Lambda)$ of smaller dimension.*

Proof: By 6.3 we know that $\pi_*(\mathrm{mic}\,\mathcal{M})$ is irreducible at its generic point. Let L be a (U_χ, K)-submodule of $\pi_*(\mathrm{mic}\,\mathcal{M})$. By 6.4 and the above remark, the support of L is Z. Thus if we take L simple, then L is unique and $(\pi_*(\mathrm{mic}\,\mathcal{M})/L) \mid \mathring{Z}$ is zero. This implies that $\mathrm{supp}(\pi_*(\mathrm{mic}\,\mathcal{M})/L)$ is contained in the closed subset $Z - \mathring{Z}$ which is of dimension smaller than Z. $\square$

Here is the main new result of this chapter:

7.2 Theorem: *If π is an isomorphism in a neighborhood of $\mathring{\Lambda}$ and if $\Gamma\mathcal{M}$ is a semisimple $(\mathfrak{g}, K)$-module, then $\Gamma\mathcal{M}$ is irreducible.*

Proof: Given an exact sequence $0 \to L \to M \to N \to 0$ of (U_χ, K)-modules, we have $\mathrm{supp}\, M = \mathrm{supp}\, L \cup \mathrm{supp}\, N$. Since $\Gamma\mathcal{M}$ is semi-simple, every submodule L of $\Gamma\mathcal{M}$ is also a quotient. By 6.4, L should have support equal to Z and by 7.1, L should have support strictly smaller than Z. Hence, either $L = \Gamma\mathcal{M}$ or $L = 0$. $\square$

7.3 Remark: The interest of this theorem is that its hypothesis are verified in many interesting cases. Indeed, the moment map π is an isomorphism in a neighborhood of $\mathring{\Lambda}$, if it is birational, and Z contains at least one point of the largest conjugacy class in $\mathcal{N}_p$, *i.e.* the Richardson class C_p. Also, if $\Gamma\mathcal{M}$ is a unitary Harish-Chandra module, then it is semi-simple. In §8, we will give a criterion for unitarity. This is the only situation to which we will apply Theorem 7.2. We stated the result in a more general framework in order to illuminate the significance of each hypothesis. We will see in theorem IV.4.5 that in this particular situation, a stronger result holds.

7.4 Example: Here is an illustrative situation which does not quite follow from theorem 7.3. I learned it from D. Vogan, and it involves discrete series representations on the symmetric space $G_r/H_r = SP_4(\mathbf{R})/SP_3(\mathbf{R})\times SP_1(\mathbf{R})$. The weight situation is the same as in I.6.7. The flag variety considered is $X = G/P$ where P is the complexification of a minimal parabolic subgroup of $SP(3,1)$; it has dimension 11. Hence the moment map is birational, see IV.4.3. There are several closed K-orbits. The one giving rise to discrete series representations is the orbit of a complex parabolic subgroup whose Levi factor intersected with $SP_4(\mathbf{R})$ is $SP_2(\mathbf{R}) \times U(1,1)$; it has dimension 5, and let us denote it by Y_1. There is another closed K-orbit – say Y_2 – which is the orbit of a complex parabolic subgroup whose Levi factor intersected with $SP_4(\mathbf{R})$ is $SP_2(\mathbf{R}) \times U(2)$; it has dimension 4. Following the notation of I.6.7, consider the weight $\lambda = (\frac{1}{2}, \frac{1}{2}, 0, 0)$, and construct the standard irreducible $(\mathcal{D}_\lambda, K)$-modules $\mathcal{M}(Y_1, \lambda)$ and $\mathcal{M}(Y_2, \lambda)$. The characteristic varieties of these modules are simply the conormal bundles of their support, and their image by the moment map π can easily be calculated. None of them contain points in the Richardson orbit of $\mathcal{N}_\mathbf{p}$, but the generic fiber of the restriction of π to $T^*_{Y_1}X$ is connected. On the other hand, the generic fiber of the restriction of π to $T^*_{Y_2}X$ consists of two disjoint copies of $\mathbf{P}^1(\mathbf{C})$. Similarly, one can check by coherent translation that the global sections of $\mathcal{M}(Y_1, \lambda)$ form an irreducible $U(\mathbf{g})$-module, while the global sections of $\mathcal{M}(Y_2, \lambda)$ break into two irreducible highest weight modules for $U(\mathbf{g})$. This shows in particular that the map $U(\mathbf{g}) \to \Gamma(X, \mathcal{D}_\lambda)$ cannot be surjective.

When $\Gamma\mathcal{M}$ is a reducible $\mathbf{g}$-module, examples of this type suggest that the number of possible submodules of
$\pi_*(\mathrm{mic}\,\mathcal{M})$ should be bounded by the number of connected components of $\pi^{-1}(z) \cap \mathring{\Lambda}$, for $z \in \mathring{Z}$. By an argument on characteristic varieties, this can be proved in special cases.

III.8 . Decomposable modules.

It can be seen with the functor L introduced in §II.4 that the global sections of standard $(\mathcal{D}, K)$-modules on the full flag variety are canonically the K-finite duals of the standard $(\mathbf{g}, K)$-modules constructed in [Vogan 2, §6.5] by cohomological parabolic induction. In this section, we will review Vogan's result which says that cohomological induction from a θ-stable parabolic subgroup preserves unitarity; hence, the $(\mathbf{g}, K)$-modules induced in this fashion from a unitary representation are decomposable, i.e. they split into a direct sum of submodules. The θ-stable assumption can be relaxed to the hypothesis that the K-orbit of the parabolic subgroup is closed, but we will not need this generality here. Moreover, for the study of $\mathcal{D}$-modules, it suffices to know the result for induction from 1-dimensional representations.

Let Z be a closed K-orbit in the flag space $\mathcal{P}$ of type P. We have a smooth fibration $\tau : \mathcal{B} \to \mathcal{P}$. $\pi^{-1}Z$ is a K-stable closed subvariety of $\mathcal{B}$; thus it is the closure of one K-orbit, say Y. Observe that $\bar{Y} = \pi^{-1}Z$ is smooth by construction.

$$\begin{array}{ccccc} Y & \hookrightarrow & \overline{Y} & \to & \mathcal{B} \\ & & \downarrow & & \downarrow \\ & & Z & \to & \mathcal{P} \end{array}$$

Let θ is the involution of G defining K. We have seen in II.1.4 that the K-orbit of $P \in \mathcal{P}$ is closed if and only if P contained a θ-stable Borel subgroup. Of course, if P itself is θ-stable, then it contains a θ-stable Borel subgroup. We are interested in a special type of parabolic subgroups P constructed as follows. Suppose that $\lambda \in \mathbf{g}^*$ is a linear form fixed by θ and purely imaginary with respect to the real form G_o (such that $K \cap G_o$ is a maximal compact subgroup of G_o). Let $L(\lambda) = \text{Cent}(\lambda; G)$ and $N(\lambda)$ be a θ-stable unipotent subgroup of G such that $P(\lambda) = L(\lambda).N(\lambda)$ is parabolic. Then $P(\lambda)$ is clearly θ-stable. For the rest of this chapter, we assume that P is of this form. Let ρ_p be the half sum of the roots in the unipotent radical N of P. Suppose $\lambda + \rho_p$ is the differential of a character of P. Consider the associated line bundle $\mathcal{O}_\lambda$ on $\mathcal{P}$. Let $i_z : Z \hookrightarrow \mathcal{P}$ be the inclusion, and let $c = codim_{\mathcal{P}} Z$. Then the global sections of the standard $(\mathcal{D}_\lambda, K)$-module $i_{z*}i_z^!\mathcal{O}_\lambda$ form the space $H_Z^c(\mathcal{P}, \mathcal{O}_\lambda)$. We regard it as a $(\mathbf{g}, K)$-module.

One can also view this module on the full flag variety $\mathcal{B}$. Denote by $\tilde{\mathcal{O}}_\lambda$ the pull back of $\mathcal{O}_\lambda$ from $\mathcal{P}$ to $\mathcal{B}$, and let $i : Y \hookrightarrow \mathcal{B}$ be the inclusion. Since we have $c = \mathrm{codim}_{\mathcal{B}} Y$, then the global sections of the standard $(\mathcal{D}_{\lambda+\rho_\ell}, K)$-module $i_* i^! \tilde{\mathcal{O}}_\lambda$ form the space $H^c_Y(\mathcal{B}, \tilde{\mathcal{O}}_\lambda)$. By base change, we have $H^c_Z(\mathcal{P}, \mathcal{O}_\lambda) = H^c_Y(\mathcal{B}, \tilde{\mathcal{O}}_\lambda)$.

Now, [Zuckerman] has defined certain $(\mathbf{g}, K)$-module denoted $A_{\mathbf{p}}(\lambda)$, see definition below. However, he used a choice of polarization opposite to ours; this has the effect of replacing the parabolic subalgebra $\mathbf{p}$ by its opposite $\overline{\mathbf{p}}$. Note that if $\mathbf{p}$ is a parabolic subalgebra which is minimal in some real form of $\mathbf{g}$, then it is conjugate by G to its opposite $\overline{\mathbf{p}}$.

8.1. Lemma: *For λ dominant in $\mathbf{t}^*_P$, we have $H^c_Z(\mathcal{P}, \mathcal{O}_\lambda) \simeq A_{\overline{\mathbf{p}}}(\lambda)$ as $(\mathbf{g}, K)$ modules.*

Proof: Recall the definitions of the functors L, Γ, *ind* and *pro* in II.4. Put $L = L^K_{L\cap K}$ and $\Gamma = \Gamma^K_{L\cap K}$. By definition, we have:

$$A_{\overline{\mathbf{p}}}(\lambda) = \Gamma^c \mathrm{pro}^{(\mathbf{g}, L\cap K)}_{(\overline{\mathbf{p}}, L\cap K)}(\mathbf{C}_\lambda \otimes \mathbf{C}_{\rho_{\overline{p}}}) .$$

One can prove – as in [Vogan 2, Prop. 8.2.15] – that all other derived functors of Γ vanish, thanks to the dominance of λ. By proposition II.4.1, the left derived functors of L also vanish in all degrees but one. Therefore we can omit superscript. The definitions of the functors in II.4 imply readily:

$$A_{\overline{\mathbf{p}}}(\lambda) \simeq L\, \mathrm{ind}^{(\mathbf{g}, L\cap K)}_{(\mathbf{p}, L\cap K)}(\mathbf{C}_\lambda \otimes \mathbf{C}_{\rho_p}) ,$$

Thanks to the dominance of λ, theorem II.4.4 implies:

$$\begin{aligned} \Delta_\lambda A_{\overline{\mathbf{p}}}(\lambda) &= \Delta_\lambda L \mathrm{ind}^{(\mathbf{g}, L\cap K)}_{(\mathbf{p}, L\cap K)}(\mathbf{C}_\lambda \otimes \mathbf{C}_{\rho_p}) \\ &= \mathcal{L} \Delta_\lambda \mathrm{ind}^{(\mathbf{g}, L\cap K)}_{(\mathbf{p}, L\cap K)}(\mathbf{C}_\lambda \otimes \mathbf{C}_{\rho_p}) . \end{aligned}$$

Consider $(\mathcal{D}, L\cap K)$-module $\mathbf{C}_{\lambda+\rho_p}$ over the point P and let $i_p : \{P\} \hookrightarrow \mathcal{P}$ be the inclusion. Then

$$\begin{aligned} \Delta_\lambda(\mathbf{C}_{\lambda+\rho_p}) &= \Delta_\lambda(U(\mathbf{g}) \bigotimes_{\mathbf{p}} \mathbf{C}_{\lambda+\rho_p}) \\ &= i_{p*} \mathbf{C}_{\lambda+\rho_p} \\ \Delta_\lambda A_{\overline{\mathbf{p}}}(\lambda) &= \mathcal{L}\, i_{p*}(\mathbf{C}_{\lambda+\rho_p}) . \end{aligned}$$

Since in our case $\mathcal{O}_\lambda$ is a well defined G-homogeneous line bundle on all of $\mathcal{P}$, then

$$\begin{aligned} \mathcal{L}\, i_{p*}(\mathbf{C}_{\lambda+\rho_p}) &= a_* q_+ pr^\circ i_{p*}(\mathbf{C}_{\lambda+\rho_p}) \\ &= i_{z*}(\mathcal{O}_\lambda \mid Z) \end{aligned}$$

where

$$\begin{array}{ccccc} K \times \mathcal{P} & \xrightarrow{\ \mathrm{q}\ } & K \underset{L\cap K}{\times} \mathcal{P} & \xrightarrow{\ \mathrm{a}\ } & \mathcal{P} \\ \mathrm{pr}\downarrow & & & & \uparrow i_z \\ \mathcal{P} & \xleftarrow{\ i_p\ } & \{P\} & \hookrightarrow & K \cdot P = Z \end{array}$$

Thus $\Delta_\lambda A_{\overline{\mathbf{p}}}(\lambda) = i_{z*}(\mathcal{O}_\lambda \mid Z) = \mathcal{H}^c_Z(\mathcal{P}, \mathcal{O}_\lambda)$. Applying the global section functor we obtain the result. □

Now let $\mathbf{h}$ be a Cartan subalgebra of $\mathbf{p}$ such that $\lambda \in \mathbf{h}^*$. Let $R(\mathbf{n}, \mathbf{h})$ be the roots of $\mathbf{h}$ in $\mathbf{g}/\mathbf{p}$. With our normalizations, the infinitesimal character of $A_{\overline{\mathbf{p}}}(\lambda)$ is $\lambda + \rho_\ell$. Moreover a weight λ is dominant in $\mathbf{t}^*_{\mathbf{P}}$, if $Re\langle\alpha, \lambda\rangle \geq 0$ for all $\alpha \in R(\mathbf{n}, \mathbf{h})$. Recall that ρ_p is half the sum of the roots in $R(\mathbf{n}, \mathbf{h})$.

8.3. Theorem: (Vogan) *If λ is dominant in $\mathbf{t}^*_{\mathbf{P}}$, and if $C_{\lambda+\rho_p}$ is a unitary representation of $(\ell, L \cap K)$, then $A_{\overline{\mathbf{p}}}(\lambda)$ is a unitary* $(\mathbf{g}, K)$*-module.*

This result is a combination of 7.1 (e) and 8.5 in [Vogan 3]. In fact the final remark in Vogan's article shows that in some cases unitarity is still true in some strip "below 0". For instance, if $\mathcal{P} = \mathbf{P}^n$ and $G = SL_{n+1}(\mathbf{C})$, $\mathbf{t}^*_{\mathbf{p}} \simeq \mathbf{C}$, and the unique root α in $R(\mathbf{n}, \mathbf{t}_p)$ is $n+1$, so that $2\rho_p = n(n+1)$. Then for $\lambda \in \mathbf{C}$ such that $Re\lambda > -n$, the above unitarity result is still true, and in fact the vanishing theorem I.6.3. also holds.

8.4. Corollary: (a) *If $Re\langle\alpha, \lambda + \rho_\ell\rangle \geq 0$ for all $\alpha \in R(\mathbf{n}, \mathbf{h})$, and if $C_{\lambda+\rho_p}$ is a unitary $(\ell, L \cap K)$-module, then $A_{\overline{\mathbf{p}}}(\lambda)$ is an irreducible unitary* $(\mathbf{g}, K)$*-module.*
(b) If $Re\langle\alpha, \lambda\rangle \geq 0$ for all $\alpha \in R(\mathbf{n}, \mathbf{h})$, and if $C_{\lambda+\rho_p}$ is a unitary representation of $(\ell, L \cap K)$, then $\Gamma(\mathcal{P}, i_{z*}\mathcal{O}_\lambda)$ is a completely reducible $(\mathbf{g}, K)$*-module.*

Proof: (b) follows from 8.2 and 8.3. (a) follows from 6.2(b), 8.2 and 8.3., using the fact that $i_{z*}\mathcal{O}_\lambda$ is an irreducible $(\mathcal{D}_{\lambda+\rho_p}), K)$ module. □

Observe that $\mathrm{char}\,\mathcal{M} \supseteq \mathrm{char}(\Delta\Gamma\mathcal{M})$, and they may be different. We obtain an interesting consequence of Theorem 6.4.

8.5. Corollary: *Suppose that $\lambda \in t^*_p$ is P-dominant. If a (g, K)-module of type $A_{\overline{\mathbf{p}}}(\lambda)$ is reducible, then all its composition factors have the same associated variety.*

Proof: Use Remark 8.2, Theorem 6.4 and Corollary 8.4(b). □

A user guide to the results of this chapter is now the following. Suppose $\mathcal{M}$ is a $(\mathcal{D}_\lambda, K)$-module on $\mathcal{P}$ whose global sections M form a decomposable $(\mathfrak{g}, K)$-module. To decide whether M is an irreducible $(\mathfrak{g}, K)$-module or not, one can proceed as follows:

1. If $\mathcal{N}_p$ is normal or if $\lambda + \rho_\ell$ is dominant, then 6.2 applies and gives a positive answer.

2. If $\mathcal{N}_p$ is not normal, we look at the characteristic variety of $\mathcal{M}$. If it is mapped birationally by π onto its image, and if its image contains some normal points of $\mathcal{N}_p$, then 7.2 applies and gives a positive answer.

3. If $\pi(\mathrm{Char}\,\mathcal{M})$ does not contain any normal point of $\mathcal{N}_p$, we compare the characteristic cycles of $\mathcal{M}$ and M. If the coefficients of the cycle of largest dimension are equal, then we can again conclude postively.

In practice, Step 3 is hard to perform. For instance let $\mathcal{M} = i_{Y*}i^!\mathcal{O}(\lambda)$ be a standard $(\mathcal{D}_\lambda, K)$-module living on a closed K-orbit Y imbedded in $\mathcal{P}$ by the map i_Y. Then the characteristic cycle of $\mathcal{M}$ is simply $[\Lambda] := [T^*_Y X]$, because $\mathcal{M}$ is irreducible. The characteristic cycle of M is the direct image of $[\Lambda]$ considered as a class in K-theory. So we can write

$$[\mathrm{char}\, M] = \pi_*([\Lambda]) = \mathcal{X}(\mathcal{P}_x, \mathcal{O}_Y(\lambda))[\pi(\Lambda)] + \text{lower dimensional terms.}$$

Here $\mathcal{X}(\mathcal{P}_x, \mathcal{O}_Y(\lambda))$ is the Euler characteristic of the fiber $\mathcal{P}_x = \pi^{-1}(x)$ and x is generic in $\pi(\Lambda)$. If $\mathcal{X}(\mathcal{P}_x, \mathcal{O}_Y(\lambda))$ equals 1 – as is the case if $\mathcal{P}_x$ is a point – then M is irreducible at its generic point, and hence by 7.2, M is an irreducible $(\mathfrak{g}, K)$-module. Indeed, the coefficient of the cycle of largest dimension in char M bounds the number of constituents of M whose characteristic variety is this cycle.

IV Singularities and Multiplicities

In III.7, we saw that the properties of the moment map π lead to results about the irreducibility of D_λ-modules considered as $\mathbf{g}$-modules. In this chapter, we continue in this direction to prove our main result, namely, that the discrete series of a reductive symmetric space is multiplicity free, except possibly for a few excetional cases.

We first state some known results about the singularities of the image of π. According to III.6.2, if the moment map is birational and has a normal image, then the global section functor Γ sends an irreducible $\mathcal{D}_\lambda$-module $\mathcal{M}$ to an irreducible $\mathbf{g}$-module or to zero, when $\lambda \in t_P^*$ is P-dominant. The study of unibranchness suggests that it should suffice to assume that π is birational and that an open dense subset of $\pi(\operatorname{char}\mathcal{M})$ is unibranch. We cannot prove this assertion in general, but we shall at least give a criterion for unibranchness in terms of multiplicities of Weyl group representations.

Then we concentrate on the multiplicity one theorem for symmetric spaces. For classical groups, it is obtained after a few computations in §4. For exceptional groups, it requires a more elaborate analysis of the endomorphisms of $C^\infty(G_o/H_o)$ which commute with the action of G_o. This is done in §5.

In section 6, we determine the multiplicities of principal series representations on a symmetric space in the generic range; they are usually larger than one. Our description shows that the multiplicity is in general the number of real orbits contained in a single complex orbit.

Finally in section 7, we present some results on the τ-invariant of representations which occur as discrete series of exceptional symmetric spaces.

IV.1 Normality

First, we recall a proposition due to Springer for (a) and Richardson for (b).

1.1. Proposition: *Let X be a complex flag space of type P for the connected reductive algebraic group G and let $\pi : T^*X \to \mathbf{g}^*$ be the moment map*

(a) *π is a proper map and* $\operatorname{Im}\pi$ *is the closure of the Richardson nilpotent conjugacy class C_p of P in $\mathbf{g}^*$,*

(b) *π is finite over C_p of degree* $b_P = \mid \operatorname{Cent}(x,G)/\operatorname{Cent}(x,P) \mid \forall x \in C_p$.

See [Steinberg, 4.2].

In particular, since T^*X is smooth, when $b_P = 1$, then π is a resolution of the singularities of its image. If B is a Borel subgroup, then X is the full flag variety, $\operatorname{Im}\pi$ is the whole nilpotent cone $\mathcal{N}^*$ in $\mathbf{g}^*$, and $b_P = 1$. Moreover, Kostant has proved that $\mathcal{N}^*$ is normal by showing that $\mathcal{N}^*$ is a local complete intersection, smooth in codimension 1. In fact, he showed that $\mathcal{N}^*$ is even a complete intersection whose defining ideal is the set of G-invariant polynomials on $\mathbf{g}^*$ without constant term. (However, it may have many singular points besides the origin.)

To study the nilpotent element of $\mathbf{g}^*$, we may assume that of $\mathbf{g}^*$ is semisimple and also we can identify $\mathbf{g}$ with $\mathbf{g}^*$ via the Killing form. The results that we shall review could be formulated for the unipotent elements of the group itself, since the exponential map – over the complex numbers – is a diffeomorphism between the set of nilpotent elements in $\mathbf{g}$ and the set of unipotent elements. A nilpotent variety is the closure of the adjoint orbit of any nilpotent element in $\mathbf{g}$. In this section, we shall denote the full flag variety by $\mathcal{B}$ and the variety of flags of type P by $\mathcal{P}$. The moment map π sends $T^*\mathcal{B}$ into $\mathcal{N}$ and $T^*\mathcal{P}$ into $\mathcal{N}_p$. Let us put $\mathcal{P}_x = \pi^{-1}(x)$ for $x \in \mathcal{N}_p$. The fiber can be described easily:

$$\mathcal{P}_x = \{\mathbf{p} \in \mathcal{P} \mid x \in \mathbf{n}(\mathbf{p})\}$$

where $\mathbf{n}(\mathbf{p})$ is the nilpotent radical of $\mathbf{p}$. $\mathcal{P}_x$ are called Spaltenstein varieties. Unless $\mathcal{P} = \mathcal{B}$, the varieties $\mathcal{P}_x$ are different from $\mathcal{P}'_x$ the fixed point set of the flow generated by x on $\mathcal{P}$:

$$\mathcal{P}'_x = \{\mathbf{p} \in \mathcal{P} \mid x \in \mathbf{p}\} .$$

Both are projective varieties, but the following result is false for $\mathcal{P}_x$ in general.

1.2. Lemma: *The varieties $\mathcal{P}'_x$ are connected.*

In fact, as we see from 1.1(b), when x is a Richardson element for $\mathbf{p}$, $\mathcal{P}_x$ consists of b_P points.

Now we will investigate the normality of a nilpotent variety, i.e. the closure of a nilpotent conjugacy class. Since a product variety is normal if and only if every factor is normal, we may assume that $\mathbf{g}$ is a simple Lie algebra.

1.3. Theorem: [Kraft-Procesi] *If $\mathbf{g} = sl_n$, then every nilpotent variety is normal. If $\mathbf{g} = sp_n$ or o_n, then every nilpotent variety is semi-normal and it is normal if and only if it is normal in codimension 2.*

By $\mathbf{g} = o_n$ is meant $\mathbf{g} = so_n$ but $G = O_n$. The fact that conjugation is taken under the full orthogonal group O_n affects only the "very even" classes which appear when n is a multiple of 4. A variety X is semi-normal if every homeomorphism $Z \to X$ is an isomorphism. So roughly it means that the nilpotent varieties in the classical groups have no cusp singularities. In fact Kraft and Procesi have determined that the only type of abnormal singularity which can appear is the product of an affine space with two singularities of type A_n glued together at one point.

IV.2 Unibranchness

A point x of a variety X is called unibranch if the preimage of x in the normalization of X is connected. A typical example of singular unibranch point is a cusp, for instance the mid-point of a curly brace $\{$; a typical example of non-unibranch point is a double point, for instance the mid-point of a plus sign $+$. A unibranch point is also called topologically normal. If an algebraic variety X is seminormal and all its points are unibranch, then X is normal.

Normality (and seminormality) is a local, open condition. Hence the set of abnormal points is always closed. But the set of multibranch points need not be closed. In the picture below, all points on the z-axis are abnormal, but 0 is unibranch.

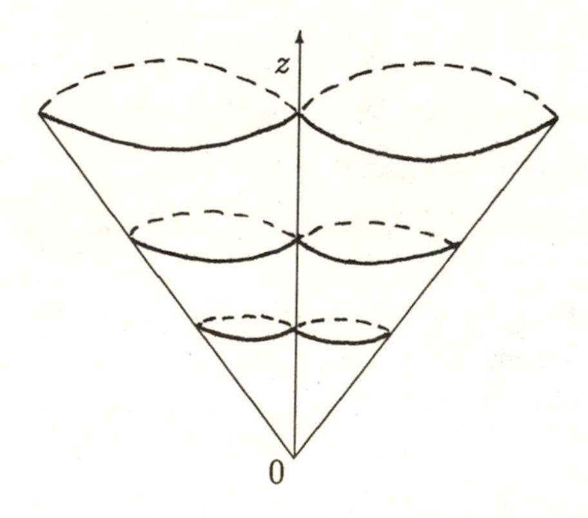

Let us denote by $IH_x^n(X)$ the nth local intersection cohomology group of X at x with coefficients in $\mathbf{Z}$. An equivalent way to express the unibranchness of a point x in a complex variety X is to ask that $IH_x^0(X) = \mathbf{C}$. Assume that the moment map $\pi : T^*\mathcal{P} \to \mathcal{N}_p$ is birational, where $\mathcal{N}_p$, is the nilpotent variety determined by the Richardson class of P. From the theory of Borho and MacPherson, the dimension of $IH_x^0(C)$ should be expressible in terms of multiplicities of Weyl group representations. We describe the answer which is quite elegant but unfortunately still hard to compute. It generalizes to arbitrary G a formula in [Borho-MacPherson, p. 710] proved for SL_n.

All conjugacy classes in $\mathbf{g}$ have even dimensions (over C). For $x \in \mathcal{N}$, let C_x be the conjugacy class of x and $2d_x = \operatorname{codim}(C_x : \mathcal{N})$. One has $d_x = \dim \mathcal{B}_x$. Let d_P be the codimension in $\mathcal{N}$ of C_p. Then $\dim \mathcal{P}_x \leq d_x - d_p$. Let σ_p be the special representation of the Weyl group W of G associated to the Richardson class C_p by Springer's correspondence. For any $x \in \mathcal{N}$, W acts on $H^i(\mathcal{B}_x)$.

2.1. Propositions: *When π is birational, $x \in \mathcal{N}_p$, we have* $\dim H^\circ(\mathcal{P}_x) = \dim IH_x^\circ(\mathcal{N}_p) = \operatorname{mtp}(\sigma_p, H^{2d_p}(\mathcal{B}_x))$.

Proof: Let $A^\bullet = R^\bullet\pi_*\mathbf{Q}_{T^*\mathcal{P}}$. If ℓ is an irreducible representation of $\tau^{-1}C_x$, $\mathbf{x} \in \mathcal{N}_p$, let $V_{(x,\varphi)}$ be a $\mathbf{Q}$-vector space of dimension equal to the multiplicity of φ in the monodromy representation of $\tau_1 C_x$ on the top cohomology group of $\mathcal{P}_x$. Let $i^* : C_x \hookrightarrow \mathcal{N}_p$ be the inclusion, and L_φ be the local system on C_x with monodromy representation φ. We shall say that the pair (C_x, φ) is relevant to π if $\dim \mathcal{P}_x = d_x - d_p$ and $V_{(x,\varphi)} \neq 0$. Since π is a semismall map – i.e. $\dim(\text{fiber}) \leq \frac{1}{2}\dim(\text{stratum})$ – we can apply a result of Borho-MacPherson to deduce for $x \in \mathcal{N}_p$

$$H^i(\mathcal{P}_x) = \bigotimes_{(x,\varphi)} IH_x^{i-2(d_x-d_p)}(\overline{C}_x; L_\varphi) \otimes V_{(x,\varphi)} \tag{2}$$

where the sum runs over the pairs (C_x, φ) relevant to π and such that $x \in \overline{C}_x \subseteq \mathcal{N}_p$.

A similar formula holds on the full flag variety

$$H^i(\mathcal{B}_x) = \bigotimes_{(x,\varphi)} IH_2^{i-2d_x}(\overline{C}_x; L_\varphi) \otimes V'_{x,\varphi}\,.$$

But here the Weyl group W acts on both sides. Borho and MacPherson have shown that the right-hand side is the decomposition of $H^i(\mathcal{B}_x)$ into irreducible representations: W acts on $V_{(x,\varphi)}$ by an irreducible representation $\sigma_{x,\varphi}$ and

$$\text{(3)} \qquad \dim IH^{i-2d_x}(\overline{C}_x, L_\varphi) = mtp(\sigma_{(x,\varphi)}, H^i(\mathcal{B}_x)) \,.$$

In this way, one obtains a bijection called the Springer correspondence:

$$\{(x,\varphi) \mid x \in \mathcal{N}/G \,,\, \varphi \in (\tau_1 C_x)^\wedge \,,\, \varphi \text{ appears in } H^{2d_x}(\mathcal{B}_x)\} \\ \leftrightarrow \{(\sigma_{(x,\varphi)}\} = W^\wedge \,.$$

If we plug formula (2) into (1), we obtain

$$\text{(4)} \qquad \dim H^i(\mathcal{P}_z) = \sum_{(x,\phi)} mtp(\sigma_{(x,\varphi)}, H^{i+2d_p}(\mathcal{B})) \cdot \dim V_{(x,\varphi)} \,,$$

$$\text{(5)} \qquad H^\circ(\mathcal{P}_z) = \bigotimes_{\substack{(x,\varphi) \\ x \in \overline{C}_n \subseteq \mathcal{N}_p}} IH_x^{-2(d_x - d_p)}(\overline{C}_x, L_\varphi) \otimes V_{(x,\varphi)} \,.$$

Since $\overline{C}_x \subset \overline{C}_p$, we have $d_x \geq d_p$. Hence for $-2(d_x - d_p)$ to be non-negative, we must have $d_x = d_p$ and so $x \in C_p$. On the other hand, for relevant (x,φ):

$$\dim V_{(x,\varphi)} = mtp(\varphi, H^{2(d_x - d_p)}(\mathcal{P}_x)) \,.$$

When $x \in C_p$, $\mathcal{P}_x$ is a point by the birationality of π, therefore

$$\begin{aligned} \dim V_{(x,\varphi)} &= 1 \quad \text{if } \varphi = \mathbb{1} \\ &= 0 \quad \text{otherwise} \,. \end{aligned}$$

Collecting this information, and since $\sigma_p = \sigma_{(x,1)}$, (3) and (4) become respectively

$$\begin{aligned} \dim H^\circ(\mathcal{P}_x) &= mtp(\sigma_p, H^{2d_p}(\mathcal{B}_x)) \\ H^\circ(\mathcal{P}_x) &= IH_z^\circ(\mathcal{N}_p) \,. \; \square \end{aligned}$$

Note two particular cases of this proposition:

- if $x \in C_p$, then x is a normal point of $\mathbf{n}_p$, hence $H^\circ(\mathbf{P}^x) = C \cdot H^{2d_p}(\mathcal{B}^x)$ is the top cohomology group of $(\mathcal{B}_x)$, and the identity 2.1 reduces to Springer's results: the special Weyl group representation σ_p occurs with multiplicity one in $H^{2d_p}(\mathcal{B}_x)$.

- if $z = 0$, then 0 is unibranch because $\pi^{-1}(0) = \mathcal{P}$. We have $\mathcal{B}_0 = \mathcal{B}$ and $H^*(\mathcal{B})$ is equivalent to the regular representation of W. The identity 2.1 reduces to Joseph's result: σ_p occurs with multiplicity one in the space of W-harmonic polynomials on t^* of degree d_p.

This criterion for unibranchness has been implemented on a computer by [Beynon–Spaltenstein]; they have obtained a complete list of the nilpotent orbits whose closure is not unibranch, for the exceptional groups of type E_n. On the other hand, [Richardson] has detected non-normal nilpotent varieties in the exceptional groups using other methods. Their combined results show the existence of nilpotent varieties which are unibranch but not normal; a phenomenon which cannot occur in the classical groups since all nilpotent varieties there are seminormal, *cf.* 1.3.

IV.3 Invariant differential operators on a reductive symmetric space

The purpose of the following three sections is to study the discrete series representations which occur in the square integrable functions on a reductive symmetric space. We will show that they are inequivalent to each other, except possibly in a few exceptional cases. In II.7.6, we saw that such a multiplicity one property holds within each eigenspace of the algebra of invariant differential operators provided the eigencharacter is sufficiently regular. Here, we shall show that under the same assumption of regularity, these eigenspaces have in fact different infinitesimal characters, hence they cannot have any common constituent. Since we shall be working henceforth on a particular type of flag varieties, we recall once more the analytic set-up and we simplify the notation slightly. This will benefit the reader who joins us at this stage.

Let G be a complex connected reductive linear algebraic group. Let σ be an involution of G with fixed point subgroup H, so that G/H is the complexification of the symmetric space we want to study. Take another involution θ of G commuting with σ, with fixed point subgroup K; θ determines which real form G_o/H_o we are studying in the sense that K is the complexification of a maximal compact subgroup in a unique – up to isomorphism – real form G_o of G. Let P be a parabolic subgroup of G which is minimal with respect to H: to find such a subgroup P, consider a σ-split torus A in G, *i.e.* a torus on which σ acts by $-id$, and suppose A is maximal for this property. Diagonalize the adjoint action of A on the Lie algebra $\mathbf{g}$ of G to obtain a root systems $R_{\mathbf{a}}$ in $\mathbf{a}$ with Weyl group $W_{\mathbf{a}}$. Then $\mathbf{g} = \mathbf{m} + \mathbf{a} + \mathbf{n} + \overline{\mathbf{n}}$ as usual where $\mathbf{m}$ is the centralizer of $\mathbf{a}$ in $\mathbf{h}$ and $\mathbf{n}$ is a nilpotent subalgebra corresponding to a choice of positive roots. Now $\mathbf{p}$ is simply $\mathbf{m} + \mathbf{a} + \overline{\mathbf{n}}$ and P is the normalizer of $\mathbf{p}$ in G.

Let X be the variety of all subgroups of G conjugate to P. X is called a flag variety of type P and it is a complete smooth projective manifold. H and K act on X: let X^0 be the unique open orbit of H and let Y denote a closed K-orbit in X which intersects (non-trivially) X^0. The stabilizer in K of any point $P \in Y$ is a connected parabolic subgrou of K. Let $T_P = P/(P,P)$ where (P,P) is the commutator subgroup of P. For different points of X, the tori T_P are canonically conjugate, so there is a well defined notion of dominance for their characters such that the positive roots at P are those which are in $\mathbf{g}/\mathbf{p}$. Define $A_P := T_P/H \cap P$. For any point $P \in X^0$, A_P is isomorphic to $A/H \cap A$ with A as above; note that $H \cap A$ is a finite abelian 2-group, therefore A and A_P have isomorphic Lie algebras. A_P can be strictly smaller for points outside of X^0. However, since the orbit Y has at least one point in X^0, it suffices to consider A_P at this point. Let ρ_p be half the sum of the roots of A_P in $\mathbf{g}/\mathbf{p}$. Take a dominant weight $\lambda \in \mathbf{a}_p^*$ such that $\lambda - \rho_p$ is a character of A_P. We suppose in addition that λ is regular, that is $\langle \lambda, \alpha \rangle \neq 0$ for all $\alpha \in R_a$. $\lambda - \rho_p$ can be viewed as a character of P; hence, it determines a line bundle $\mathcal{L}$ over the variety Y. Let $\mathcal{O}(Y, \lambda)$ be the sheaf of holomorphic sections of $\mathcal{L}$. There exists a sheaf $\mathcal{D}_\lambda$ of twisted differential operators on X and using the direct image functor associated to the inclusion map $i : Y \to X$, one obtains as in chapter II a $(\mathcal{D}_\lambda, K)$-module $\mathcal{M}(Y, \lambda)$ on X supported on Y which consists of $\mathcal{O}(Y, \lambda)$ together with all the derivatives of its local sections in the directions normal to Y. Let $M(Y, \lambda)$ denote the global sections of $\mathcal{M}(Y, \lambda)$. Let D_λ be the ring of global sections of $\mathcal{D}_\lambda$.

3.1. Theorem: *The (D_λ, K)-modules $M(Y, \lambda)$ are irreducible or zero, and for different K-orbits Y they are pairwise inequivalent.*

Theorem I.6.3.(1) and proposition I.6.6 imply the first part; the inequivalence of the $\mathcal{D}_\lambda$-modules follows from the fact that they have different characteristic varieties. Now Flensted-Jensen and Oshima-Matsuki have proved that the real symmetric space G_o/H_o corresponding to our choice of subgroup K, has a discrete series if and only if rank G/H = rank $K/H \cap K$.

This condition means that A is conjugate to a subgroup of K, and we shall make this hypothesis in the next section. Then as Y and λ run through their allowed ranges, the modules $M(Y,\lambda)$ considered as $(\mathbf{g},K)$-modules exhausts the discrete series representations of G_o/H_o.

Let $\mathbf{t}$ be a Cartan subalgebra of $\mathbf{g}$ containing $\mathbf{a}$ and let $\mathbf{b}$ be a Borel subalgebra such that $\mathbf{t} \subset \mathbf{b} \subset \mathbf{p}$. Let R be the root system of $\mathbf{t}$ in $\mathbf{g}$; the positive roots being those which are in $\mathbf{g}/\mathbf{b}$. Let ρ_ℓ be half the sum of the positive roots of $\mathbf{t}$ in $\mathbf{m}$. A better notation for ρ_ℓ would be $\rho(m)$; however, the notation ρ_ℓ has the advantage to agree with our convention of section I.3. The universal enveloping algebra U of $\mathbf{g}$ maps into D_λ. In most cases the weight $\lambda + \rho_\ell$ is dominant in $\mathbf{t}^*$; this implies that the map $U \to D_\lambda$ is surjective according to III.6.2(b).

3.2. Theorem: *If $\lambda+\rho_\ell$ is dominant in $\mathbf{t}^*$, then the $(\mathbf{g},K)$-modules $M(Y,\lambda)$ are irreducible or zero, and for different K-orbits Y they are pairwise inequivalent.*

Now we allow λ to vary. Let W (resp. W_a) be the Weyl group of R (resp. R_a); it acts naturally on $\mathbf{t}^*$ (resp. $\mathbf{a}^*$). The Harish-Chandra isomorphism identifies $Z(\mathbf{g})$ with $S(\mathbf{t})^W$: the algebra of W-invariant polynomial functions on $\mathbf{t}^*$. Hence the maximal spectrum of $Z(\mathbf{g})$ is identified with $\mathbf{t}^*/W$. The induced map $Z(\mathbf{g}) \to D_\lambda$ factors through $\mathbf{C}$: the constant differential operators on X, and it coincides with $\chi_{\lambda+\rho_\ell}$: the character associated to $\lambda + \rho_\ell$ even if this weight is not dominant.

In addition, $Z(\mathbf{g})$ considered as the algebra of bi-invariant differential operators on G maps naturally into the algebra $\mathbf{D}(G/H)$ of left invariant differential operators on G/H. For any algebra R on which the group H acts, let U^H denote the elements of U which are fixed by H. We know thanks to Harish-Chandra, that $\mathbf{D}(G/H)$ is isomorphic to $U^H/(U\cdot\mathbf{h})\cap U^H)$ where this latter algebra acts from the right on functions on G/H; recall that the group G acts by left translations. Since these operators all have polynomial coefficients, the structure of the real algebra $\mathbf{D}(G_o/H_o)$ is the same as the one of the algebra $\mathbf{D}(G/H)$. By a Chevalley type isomorphism, $\mathbf{D}(G/H)$ is canonically isomorphic to $S(\mathbf{a})^{W_a}$.

3.3 Proposition: [Helgason 1 §7, & 4] *The algebra morphism*

$$f : Z(\mathbf{g}) \to \mathbf{D}(G/H)$$

is surjective unless $\mathbf{g}/\mathbf{h}$ *contains factors of type* $E_6/\mathbf{so}(10) \oplus \mathbf{C}$, E_6/F_4, $E_7/E_6 \oplus \mathbf{C}$ *or* $E_8/E_7 \oplus \mathbf{sl}(2)$.
Furthermore, it is always surjective at the level of fraction fields.

The last statement means that for any $D \in \mathbf{D}(G/H)$, one can find two elements Z_1, $Z_2 \in Z(\mathbf{g})$, such that $f(Z_1)\cdot D = f(Z_2)$. The map corresponding to f at the level of maximal spectra is $f^{\#} : \mathbf{a}^*/W_a \to \mathbf{t}^*/W$. It is in general nonlinear although both spectra are affine spaces, but it is birational onto its image. Moreover, $f^{\#}(\lambda)$ is simply $\lambda + \rho_\ell$ if this latter weight is dominant in $\mathbf{t}^*$. We denote by ψ_λ the character of $\mathbf{D}(G/H)$ associated to a weight $\lambda \in \mathbf{a}^*$.

The modules $M(Y, \lambda)$ admit a scalar action of $\mathbf{D}(G/H)$ given by the character ψ_λ. This can only be seen using the realization of $M(Y, \lambda)$ in the space of analytic functions on the symmetric space G_o/H_o that we now recall. Consider the space $A(G_o/H_o)$ of real analytic functions on G_o/H_o. Given the character ψ_λ of $\mathbf{D}(G/H) \approx \mathbf{D}(G_o/H_o)$, we consider the corresponding eigenspace $A(G_o/H_o; \psi_\lambda)$. The G_o-module structure of this space is determined by the $(\mathbf{g}, K)$-module structure of its Harish-Chandra module $A_K(G_o/H_o; \psi_\lambda)$ which consists of all eigenfunctions transforming according to some finite dimensional representation of K_o that we can then extend canonically to an algebraic representation of K. Let $L^2_K(G_o/H_o; \psi_\lambda)$ denote the subspace of square integrable functions in $A_K(G_o/H_o; \psi_\lambda)$ with respect to a natural invariant measure. Then we have Oshima – Matsuki's result seen in II.7.4:

3.4. Theorem: *If* λ *is a dominant regular weight in* $\mathbf{a}^*$ *such that* $\lambda - \rho_\ell$ *extends to a character of* A_P, *then* $L^2_K(G_o/H_o; \psi_\lambda) \simeq \oplus_Y M(Y, \lambda)$ *where* Y *run through the closed* K*-orbits whose intersection with* X^0 *is non-empty. Otherwise,* $L^2_K(G_o/H_o; \psi_\lambda) = 0$.

Thus we have the following diagram for the action of $Z(\mathbf{g})$ on $M(Y,\lambda)$.

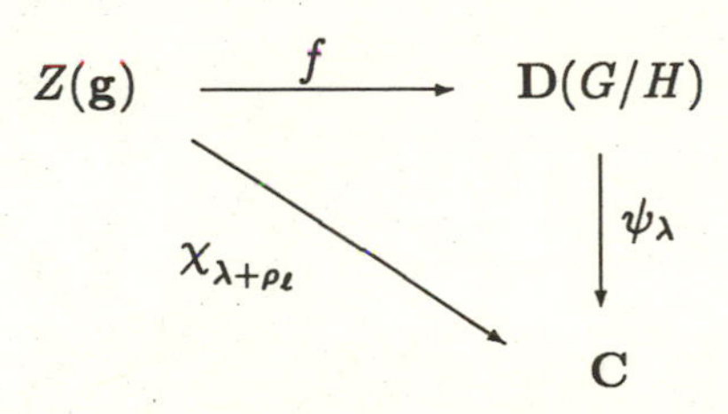

In the exceptional cases where f is not surjective, it happens that two weights $\lambda, \mu \in \mathbf{a}^*$ non-conjugate by W_a may be conjugate by W when ρ_ℓ is added to them. In other words $f^\#$ is not injective or if you prefer, although $\psi_\lambda \neq \psi_\mu$, we may have $\chi_{\lambda+\rho_\ell} = \chi_{\mu+\rho_\ell}$. If this happens, we shall say that λ (and μ) are *ambiguous*, and also that the infinitesimal character $\chi_{\lambda+\rho_\ell}$ is *ambiguous*. The following lemma shows that most weight are not ambiguous.

3.5. Lemma: *Let $\lambda, \mu \in \mathbf{a}^*$. If $\lambda + \rho_\ell$ is dominant in $\mathbf{t}^*$ and $\psi_\lambda \neq \psi_\mu$, then $\chi_{\lambda+\rho_\ell} \neq \chi_{\mu+\rho_\ell}$.*

Proof: In fact we shall prove a much more detailed statement. Thanks to Helgason's result, it suffices to examine the four exceptional cases. We shall use the canonical coordinates on $\mathbf{t}^*$ given by evaluating the weights against the simple coroots and we will draw them in the same shape as the Dynkin diagram. In each case, $\mathbf{m}$ is a simple Lie algebra of type D_4. We will represent the roots of $\mathbf{m}$ by w and the roots whose restriction to $\mathbf{a}$ is non-trivial by $\circ$, so we obtain the Satake diagram of the real form of $\mathbf{g}$ with $\mathbf{h}$ maximal compact subalgebra. Next to the Dynkin diagram of $\mathbf{g}$, we draw the restricted Dynkin diagram given by the roots of $\mathbf{a}$. The indices below the vertices are the labels of the simple roots; e.g. 1 stands for α_1.

$E_6/\mathbf{so}(10) \oplus \mathbf{C}$

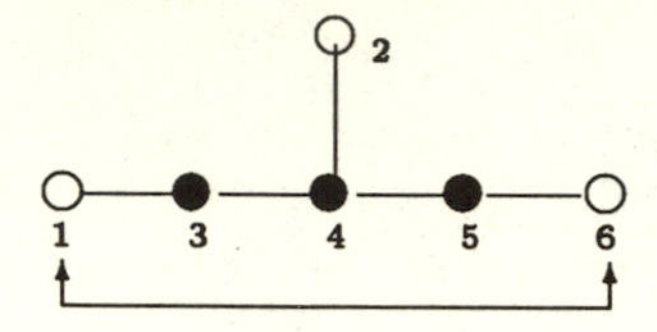

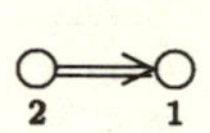

E_6/F_4

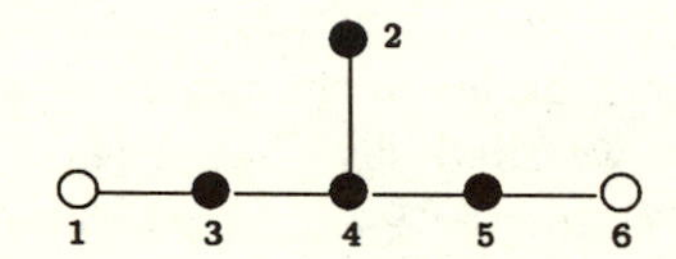

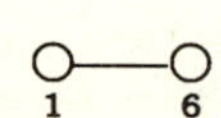

$E_7/E_6 + \mathbf{C}$

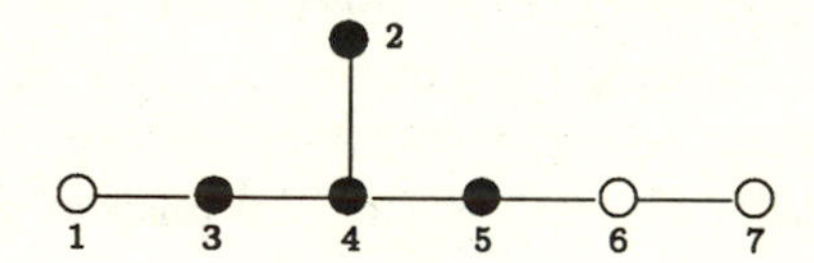

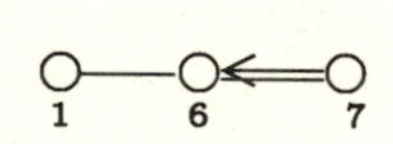

$E_8/E_7 + \mathbf{sl}(2)$

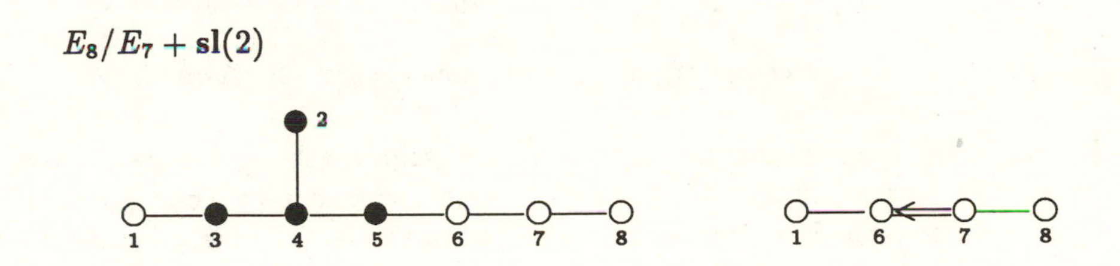

ρ_ℓ has the following coordinates:

$$\begin{matrix} & & -2 & & \\ -\frac{3}{2} & 1 & 1 & 1 & -\frac{3}{2} \end{matrix} ,$$

$$\begin{matrix} & & 1 & & \\ -3 & 1 & 1 & 1 & -3 \end{matrix} ,$$

$$\begin{matrix} & & 1 & & & \\ -3 & 1 & 1 & 1 & -3 & 0 \end{matrix} ,$$

$$\begin{matrix} & & 1 & & & & \\ -3 & 1 & 1 & 1 & -3 & 0 & 0 \end{matrix}$$

Now the weights $\lambda \in \mathbf{a}^*$ are given by their coordinates on the restricted Dynkin diagrams. A combinatorial computation shows that the only dominant weights λ, $\mu \in \mathbf{a}^*$ which are not conjugate by the small Weyl group W_a, but such that $\lambda + \rho_\ell$ is W-conjugate in $\mathbf{t}^*$ to $\mu + \rho_\ell$ are the following ones:

		λ	μ
$E_6/\mathbf{so}(10) \oplus \mathbf{C}$		$(3,\frac{1}{2})$	$(0,\frac{5}{2})$
E_6/F_4		(4,1)	(1,4)
E_7/E_6		(4,1,m)	(1,4,m-2)
	$\mathrm{Re}(m) < 2$	(4,1,m)	(1,m+2,2-m)
$E_8/E_7 + \mathbf{sl}(2)$		(4,1,m,n)	(1,4,m-2,n)

$$\begin{cases} \text{Re}(m) < 2 \\ \text{Re}(m+n) \geq 2 \end{cases} \quad (4,1,m,n) \quad (1,m+2,2-m,m+n-2)$$

$$\text{Re}(m+n) < 2 \quad (4,1,m,n) \quad (1,m+2,n,2-m-n)$$

In each case, m and n can be any complex numbers with non-negative real parts. The lemma follows from this list of weights since none of the ambiguous weights satisfies the hypothesis. For example, in the E_6/F_4 case if $\lambda = (4,1)$, then

$$\lambda + \rho_\ell = \begin{matrix} & & 1 & & \\ 1 & 1 & 1 & 1 & -2 \end{matrix}$$

which is not dominant. □

It follows from this lemma and corollary II.7.6 that if V is an irreducible Hilbert space representation of G_o with infinitesimal character $\chi_{\lambda+\rho_\ell}$ for some dominant weight $\lambda \in \mathbf{a}^*$ such that $\lambda + \rho_e ll$ is also dominant in $\mathbf{t}^*$, then the dimension of $Hom_{G_o}(V, L^2(G_o/H_o))$ is at most one. In other words, we have proved the following improvement on corollary II.7.6:

3.6. Proposition: *Suppose $\lambda+\rho_\ell$ is dominant in $\mathbf{t}^*$, then the discrete series of G_o/H_o with infinitesimal character $\lambda + \rho_e ll$ is multiplicity free.*

IV.4 Multiplicity one within an eigenspace of $\mathbf{D}(G/H)$

In this section we extend Theorem 3.2 to the case where $\lambda \in \mathbf{a}^*$ is dominant but $\lambda+\rho_\ell \in \mathbf{t}^*$ is not. We only consider closed K-orbits Y whose intersection with the open H-orbit X^0 is non-empty. In more general situations, the result is not true. In each orbit Y, we choose a point $\mathbf{p} \in X^0 \cap Y$. Let $\mathbf{a}$ be a maximal abelian subspace on which σ is $-id$. We make the hypothesis $\text{rank}\, G/H = \text{rank}\, K/H \cap K$, so that we can take $\mathbf{a} \subset \mathbf{k}$. Let us fix a dominant weight $\lambda \in \mathbf{a}^*$, for which there exist standard H-spherical $(\mathcal{D}_\lambda, K)$-modules $\mathcal{M}(Y,\lambda)$. Since the K-orbit Y is closed and we are in the equal rank situation, this forces λ to be integral. One can improve theorem III.7.2 for the modules considered here.

4.1. Theorem [Vogan 4]: *The $(\mathbf{g}, K)$-modules $M(Y,\lambda)$ are irreducible.*

In fact, modulo some case by case computations, we will prove this result, and also the following one:

4.2. Theorem: *For different closed K-orbits Y, the (g, K)-modules $M(Y, \lambda)$ are pairwise inequivalent.*

4.3. Corollary: *The discrete eigenspaces of* $\mathbf{D}(G/H)$ *in* $L^2(G_o/H_o)$ *are multiplicity free.*

This corollary follows from the above result, and theorem II.7.5. Now, we start the proof of 4.1, and 4.2. We will first give the results of some computations for classical groups and G_2 which show that in many cases, the enveloping algebra U surjects onto D_λ. Then, we will describe a general argument based on microlocal methods, which applies to a few more representations. Finally, we will explain how to treat the remaining representations, using a reduction technique and the translation principle.

The cotangent space T^*X of X can be described by

$$T^*X = \{(\mathbf{p}, \xi) \in X \times \mathbf{g}^* \mid \xi \times \mathbf{p}^\perp\}$$

Let $\mathcal{N}_\mathbf{p}$ be the image of the moment map

$$\pi : T^*X \longrightarrow \mathbf{g}^* : (p, \xi) \longrightarrow \xi \, .$$

To simplify the notation, we identify $\mathbf{g}$ with $\mathbf{g}^*$ in a G-equivariant manner. Then $\mathcal{N}_\mathbf{p}$ is identified with the closure of a nilpotent conjugacy class $C_\mathbf{p}$ in $\mathbf{g}$ called the Richardson class of $\mathbf{p}$: it is the largest conjugacy class which intersects $\mathbf{n}$: the nilpotent radical of $\mathbf{p}$. Recall that the parabolic subalgebra $\mathbf{p}$ is by definition minimal for a real form G^r of G in which $H \cap G^r$ is compact.

4.4 Property: *Let $X = G/P$ be a flag variety. If the parabolic subgroup P is the complexification of a minimal parabolic subgroup in some real form of G, then the moment map π is birational.*

This follows from 1.1.(b), and [Hesselink]'s observation that under the above hypothesis, the centralizer of an element of $C_\mathbf{p}$ is contained in a conjugate of P. Nilpotent conjugacy classes are classified by weighted Dynkin diagrams, cf. [Springer-Steinberg]. The Richardson class $C_\mathbf{p}$ is the complexification of the orbit of the regular nilpotent elements in the real form of G associated to H as in II.2. Its weighted Dynkin diagram is straightforward to describe. Consider the Satake diagram of this real group. Put labels 0 on the black dots and labels 2 on the white dots. Note that these Richardson classes are always even.

Step 1. For the classical groups, we can translate these Dynkin diagrams into partitions, by applying the inverse of the rule described in [Springer-Steinberg Ch. IV]. The answer is summarized in the following table with the obvious restrictions on p and q.

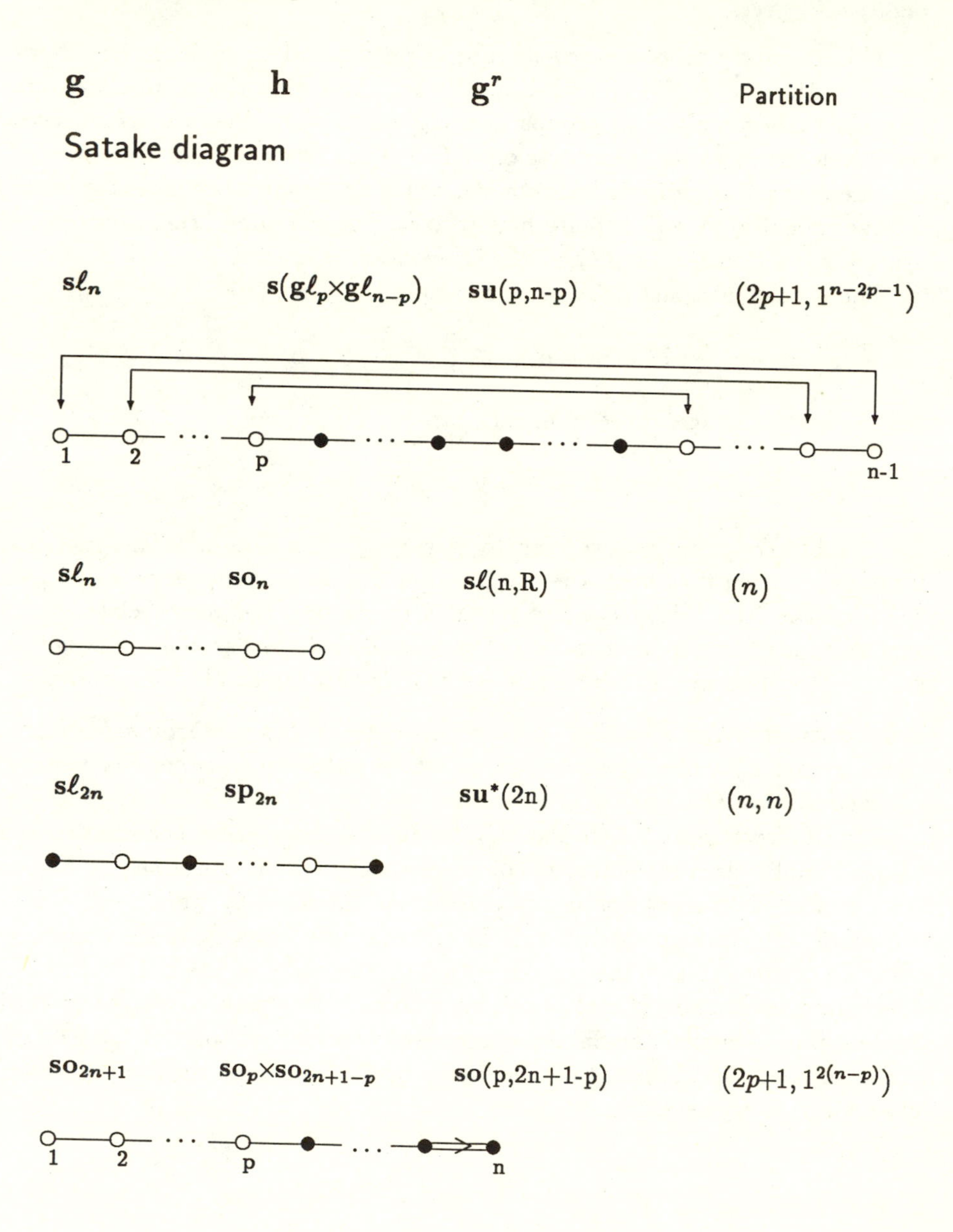

$\mathfrak{g}$	$\mathfrak{h}$	$\mathfrak{g}^r$	Partition
Satake diagram			
$\mathfrak{s}\ell_n$	$\mathfrak{s}(\mathfrak{g}\ell_p \times \mathfrak{g}\ell_{n-p})$	$\mathfrak{su}(p,n-p)$	$(2p+1, 1^{n-2p-1})$
$\mathfrak{s}\ell_n$	$\mathfrak{so}_n$	$\mathfrak{s}\ell(n,R)$	(n)
$\mathfrak{s}\ell_{2n}$	$\mathfrak{sp}_{2n}$	$\mathfrak{su}^*(2n)$	(n,n)
$\mathfrak{so}_{2n+1}$	$\mathfrak{so}_p \times \mathfrak{so}_{2n+1-p}$	$\mathfrak{so}(p,2n+1-p)$	$(2p+1, 1^{2(n-p)})$

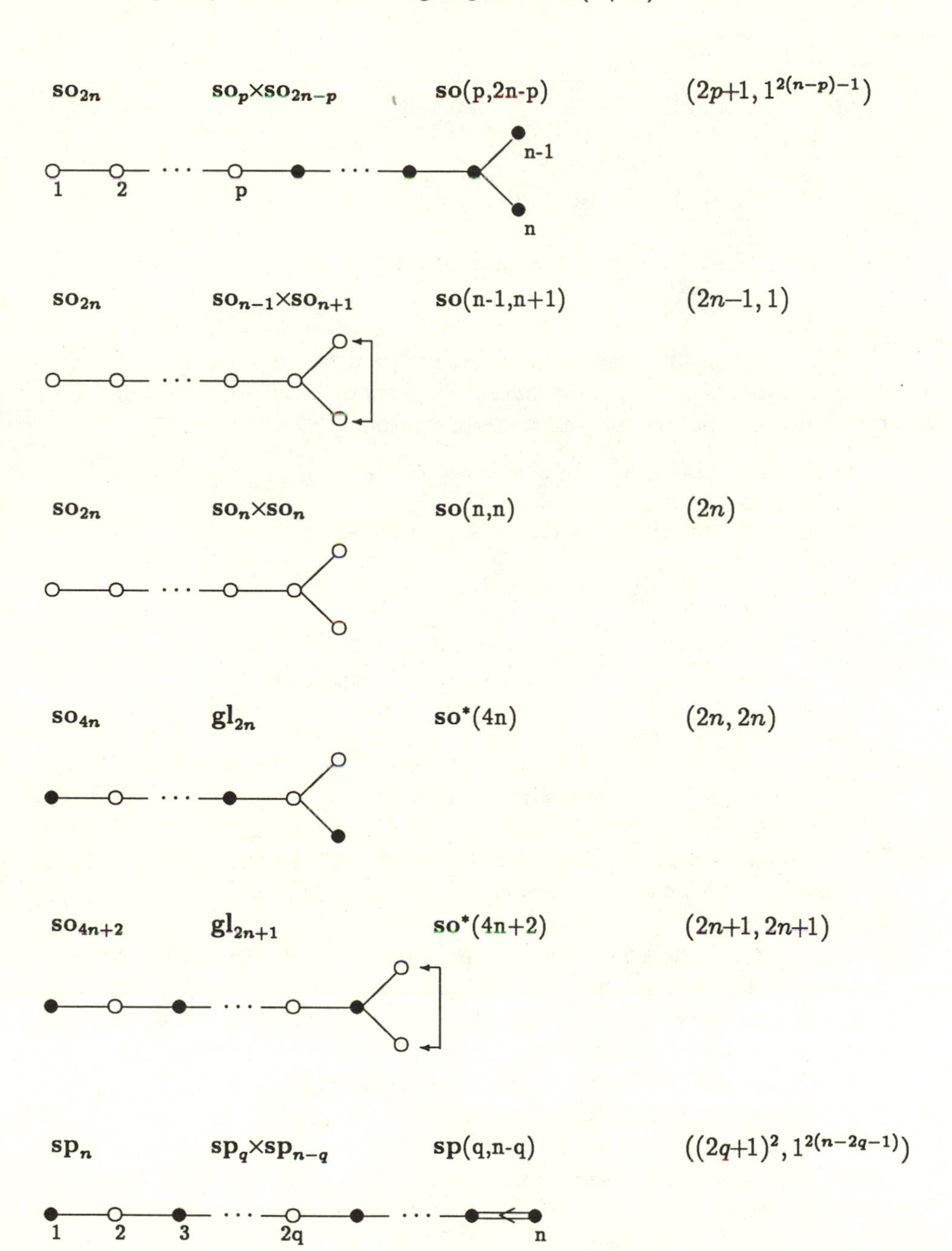
$\mathfrak{so}_{2n}$ $\mathfrak{so}_p \times \mathfrak{so}_{2n-p}$ $\mathfrak{so}(p,2n-p)$ $(2p+1, 1^{2(n-p)-1})$
1 2 p n-1 n
$\mathfrak{so}_{2n}$ $\mathfrak{so}_{n-1} \times \mathfrak{so}_{n+1}$ $\mathfrak{so}(n-1,n+1)$ $(2n-1, 1)$
$\mathfrak{so}_{2n}$ $\mathfrak{so}_n \times \mathfrak{so}_n$ $\mathfrak{so}(n,n)$ $(2n)$
$\mathfrak{so}_{4n}$ $\mathfrak{gl}_{2n}$ $\mathfrak{so}^*(4n)$ $(2n, 2n)$
$\mathfrak{so}_{4n+2}$ $\mathfrak{gl}_{2n+1}$ $\mathfrak{so}^*(4n+2)$ $(2n+1, 2n+1)$
$\mathfrak{sp}_n$ $\mathfrak{sp}_q \times \mathfrak{sp}_{n-q}$ $\mathfrak{sp}(q,n-q)$ $((2q+1)^2, 1^{2(n-2q-1)})$
1 2 3 2q n

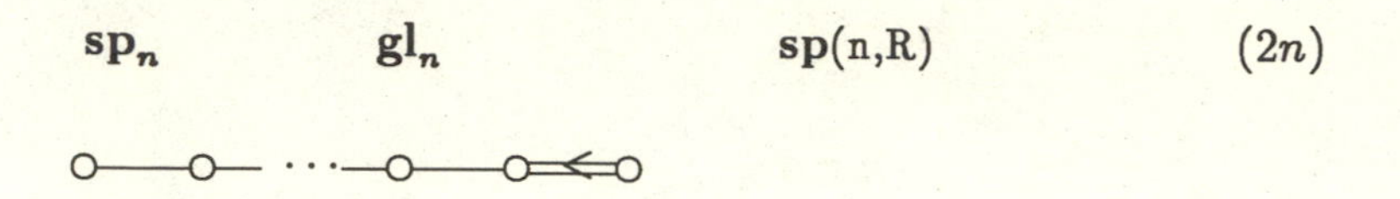

Remark: When $2p = n$ in the case of $\mathbf{su}(\mathrm{p,p})$, $\mathbf{so}(\mathrm{p,p})$, the corresponding partitions are simply (n), and if $2q = n$ in the case of $\mathbf{sp}(\mathrm{q,q})$, the partition is (n, n).

It is worth noting that the Dynkin diagram of the Levi factor L of the minimal parabolic P is simply the subdiagram of black vertices. For further information on Satake diagrams, see [Helgason 2, p. 532].

In light of [Kraft-Procesi], one draws the following conclusions:

1. if $\mathbf{g} = \mathbf{s}\ell_n$ all nilpotent varieties are normal.

2. if $\mathbf{g} = \mathbf{so}_n$, all nilpotent varieties considered here are normal.

3. if $\mathbf{g} = \mathbf{sp}_n$, the nilpotent varieties considered here are normal if and only if $\mathbf{h} = \mathbf{g}\ell_n$ or $\mathbf{sp}_q \times \mathbf{sp}_q (n = 2q)$ or $\mathbf{sp}_{q,q+1} (n = 2q + 1)$

The conjugacy class $(2n, 2n)$ in $\mathbf{so}_{4n}$ is very even. This partition describes one O_{4n}-conjugacy class which splits into two disjoint SO_{4n}-conjugacy classes. The image of the moment map is the closure of one component. Kraft and Procesi have proved [Prop. 1.5.4] that this SO_{4n}-nilpotent variety is normal. For the other cases, it suffices to apply theorem 1.3 above, and the fact that these partitions have no degenerations of type $(2n, 2n) \rightarrow (2n-1, 2n-1, 1, 1)$, cf. [Kraft-Procesi].

However, for the symplectic groups, the partition $((2q + 1)^2, 1^{2(n-2q-1)})$ shows that the corresponding conjugacy class is not normal. In codimension 2, it has a singularity which looks like two singularities of type A_q glued at one point, times an affine line. So for $q = 1$, the picture of §2 gives a heuristical idea of the kind of singularity we have in hands.

If $G = G_2$, there is only one non-trivial possibility for H, namely, $SL_2 \times SL_2$ corresponding to the split form of G_2. Then X is the full flag variety and $\mathcal{N}_p^* \simeq \bar{C}_p$ is the whole nilpotent cone which is known to be normal by Kostant's result.

As before, suppose that P is a parabolic subgroup of G, which is the complexification of a minimal parabolic subgroup in some real form G^r of G, for which $H \cap G^r$ is compact. Let X be a flag variety X of type P. Using proposition III.6.2(a), we obtain:

4.5. Proposition: *The enveloping algebra U surjects onto the ring of twisted differential operators $\Gamma(X, \mathcal{D}_\lambda)$ for all $\lambda \in \mathbf{t}_{\mathbf{p}}^*$, if $\mathbf{g} = \mathbf{s}\ell_n$, $\mathbf{so}_n$, $\mathbf{G}_2$, for all $n \in \mathbf{N}$, or if $\mathbf{g} = \mathbf{sp}_n$ with $\mathbf{g}_r = \mathbf{sp}_n(\mathbf{R})$, $\mathbf{sp}_{q,q}$ or $\mathbf{sp}_{q,q+1}$, for all $n \in \mathbf{N}$ and $n = 2q$ or $2q+1$, i.e. $\mathbf{h} = \mathbf{gl}_n$, $\mathbf{sp}_q \times \mathbf{sp}_q$, or $\mathbf{sp}_q \times \mathbf{sp}_{q+1}$.*

Now, we fix $\mathbf{g} = \mathbf{sp}_n, \mathbf{h} = \mathbf{sp}_\ell \times \mathbf{sp}_{n-\ell}$, with $\ell \leq n - l$, and we look at the real forms of the pair $(\mathbf{g}, \mathbf{h})$. The choice of the group K determines these real forms. First, we take $K = GL_n$ so that $G_r = SP_n(\mathbf{R})$ is the split real form and $H_r = SP_\ell(\mathbf{R}) \times SP_{n-\ell}(\mathbf{R})$. The rank of the symmetric space G_r/H_r is ℓ. The order of the Weyl group $W_H(A)$ is $2^\ell \cdot \ell!$. $K_o/H_o \cap K_o$ is $U(n)/SP(\ell) \cap SP(n-\ell)$, which has rank ℓ, and the order of $W_{K\cap H}(A)$ is also $2^\ell \cdot \ell!$. Recall from section II.7 that the coset space $W_H(A)/W_{K\cap H}(A)$ parametrizes the closed orbits Y supporting discrete series modules. Thus, we see that in this case there is a *unique* such orbit Y. Moreover the image $\mu(T_Y^* X) \subset \mathbf{g}^*$ contains nilpotent elements associated to the partition $((2\ell+1)^2, 1^{2(n-2\ell-2)})$; these are generic in $\mathcal{N}_p = \pi(T^*X)$, hence they are normal points of the image of π. Using Theorem III.7.2, we conclude that if $\mathcal{M}$ is a standard $(\mathcal{D}_\lambda, K)$-module supported by the closed K-orbit Y with $\lambda \in \mathbf{t}_p^*$, then $\Gamma(X, \mathcal{M})$ is an irreducible $(\mathbf{g}, K)$-module. However $U(\mathbf{g})$ does not surject onto $\Gamma(X, \mathcal{D}_\lambda)$ for certain dominant λ such that $\lambda + \rho_\ell \in \mathbf{t}^*$ is not B-dominant, *cf.* example III.7.4.

Next, we take $K = SP_k \times SP_{n-k}$. Then $G_r = SP(k, n-k)$ and $H_r = SP(r, \ell-r) \times SP(k-r, n-\ell-k+r)$ for some $r \leq k$, and $r \leq \ell$. As before, the rank of G_o/H_o is ℓ, and the order of $W_H(A)$ is $2^\ell \cdot \ell!$. Moreover, $K_o/H_o \cap K_o$ is $SP(k) \times SP(n-k)/SP(r) \times SP(k-r) \times SP(\ell - r) \times SP(n-k-\ell+r)$, which has rank $min(r, k-r) + min(\ell - r, n-k-\ell+r)$. This number can be smaller than ℓ, in which case the space $L^2(G_o/H_o)$ does not contain any discrete series representation by theorem II.7.2. When the rank of $K_o/H_o \cap K_o$ is equal to ℓ, the order of $W_{K\cap H}(A)$ is in general smaller than the order of $W_H(A)$. For instance, if $k = 2r$, then the order of $W_{K\cap H}(A)$ is equal to

$2^\ell \cdot r!\,(\ell - r)!$, and there are $\begin{pmatrix} l \\ r \end{pmatrix}$ closed K-orbits in X giving rise to discrete series representations in $L^2(G_o/H_o)$. For example, for $G_o = SP(2,2)$, and $H_o = SP(1,1) \times SP(1,1)$, there are two orbits. We will explain below how to prove that the representations coming from these two different orbits are inequivalent.

Concerning irreducibility, consider the situation $r = k \geq 2\ell$. The most regular elements of the image of the moment map $\pi(T^*_Y X)$ correspond to the partition $(3^{2\ell}, 1^{2(n-3\ell)})$. Comparing it with the partition for the Richardson class in C_p, we see that if the rank ℓ of G_o/H_o is 1, then we can conclude that irreducibility is preserved for dominant λ and $(\mathcal{D}_\lambda, K)$-modules supported on Y. In higher rank we cannot conclude by this method. However, observe that the centralizer of a generic element x of $\pi(T^*_Y X)$ is always connected. It would be interesting to know if this implies that x is unibranch, i.e. $\pi^{-1}(x)$ is connected.

Step 2. Let $\mathbf{s}$ be the (-1)-eigenspace of θ in $\mathbf{g}$; $\mathbf{s}$ is also the K-stable complement of $\mathbf{k}$ in $\mathbf{g}$. The characteristic variety of $\mathcal{M}(Y,\lambda)$ is $\Lambda := T^*_Y X$: the conormal bundle of Y in X. Its image by π is $Z := K \cdot (\mathbf{n} \cap \mathbf{s})$, where $\mathbf{n}$ is the nilpotent radical opposite to the point $\mathbf{p} \in Y$ that we chose at the beginning of this section. Z equals the associated variety $\operatorname{supp} M$ of the $(\mathbf{g}, K)$-module $M(Y,\lambda)$, whenever $M(Y,\lambda) \neq 0$. Since λ is P-dominant in $\mathbf{a}^*$, by the remark after theorem III.7.2, we see that a sufficient condition for $M(Y,\lambda)$ to be non-zero, is that $\mathbf{n} \cap \mathbf{s} \cap C_p$ be non-empty.

Recall that in chapter III, we have constructed micro-localizations $\mathcal{E}_\lambda$ for D_λ, $\mathcal{U}_\chi$ for U_χ, $\operatorname{mic} \mathcal{M}(Y,\lambda)$ for $M(Y,\lambda)$ and $\operatorname{mic} M(Y,\lambda)$ for $M(Y,\lambda)$ in such a way that $\operatorname{mic} \mathcal{M}(Y,\lambda)$ is an $\mathcal{E}_\lambda$-module and $\operatorname{mic} M(Y,\lambda)$ is a $\mathcal{U}_\chi$-module. All these new objects are sheaves on T^*X or $\mathcal{N}_p$. If we have two distinct K-orbits Y_1 and Y_2, with base points $\mathbf{p}_1$, $\mathbf{p}_2$ respectively, giving rise to two irreducible $\mathcal{D}_\lambda$-modules $\mathcal{M}(Y_1,\lambda)$ and $\mathcal{M}(Y_2,\lambda)$, then there is no (non-zero) map between the irreducible $\mathcal{E}_\lambda$-modules $\operatorname{mic} \mathcal{M}(Y_1,\lambda)$ and $\operatorname{mic} \mathcal{M}(Y_2,\lambda)$, because they are inequivalent since they have different supports. For a scheme S, we will denote by S^0 its generic point.

4.6 Theorem: *Suppose that λ is dominant in $\mathbf{a}^*$, and that both $\mathbf{n}_1 \cap \mathbf{s} \cap C_p$ and $\mathbf{n}_2 \cap \mathbf{s} \cap C_p$ are non-empty, then the $(\mathbf{g}, K)$-modules $M(Y_1,\lambda)$ and $M(Y_2,\lambda)$, are irreducible and inequivalent.*

Proof: If $\mathbf{n}_i \cap \mathbf{s} \cap C_p$ is non-empty, for $i = 1, 2$, then it consists only of normal points of $\mathcal{N}_p$. Under this hypothesis, theorem III.7.2 implies that the modules $M(Y_1, \lambda)$ and $M(Y_2, \lambda)$ are irreducible $(\mathbf{g}, K)$-modules. (This improves lemma III.5.1.(c)).

Futhermore, we can prove they are inequivalent as follows. The set of normal points of a scheme is open, thus we can find an open neighborhood Λ' of $(\Lambda_1 \cup \Lambda_2)^0$ in T^*X which maps isomorphically onto an open neighborhood Z' of $(Z_1 \cup Z_2)^0$ in $\mathcal{N}_p$ such that Z' is contained in the set of normal points of $\mathcal{N}_p$. Since λ is P-dominant, the direct image functor π_* in the category of sheaves of rings is exact with inverse π_*, hence it yields an isomorphism between the sheaves of algebras

$$\mathcal{E}_\lambda \mid_{\Lambda'} \simeq \mathcal{U}_\chi \mid_{Z'}$$

and the sheaves of modules

$$\operatorname{mic} \mathcal{M}(Y_i, \lambda)|_{\Lambda'} \simeq \operatorname{mic} M(Y_i, \lambda)|_{Z'} .$$

If $M(Y_1, \lambda)$ were equivalent to $M(Y_2, \lambda)$ over $(\mathbf{g}, K)$, then there would be an isomorphism between $\operatorname{mic} M(Y_1, \lambda)$ and $\operatorname{mic} M(Y_2, \lambda)$ over $\mathcal{U}_\chi$ because microlocalization is an exact and faithful functor. Hence there would be an $\mathcal{E}_\lambda|_{\Lambda'}$-isomorphism between the sheaves $\operatorname{mic} \mathcal{M}(Y_1, \lambda)|_{\Lambda'}$ and $\operatorname{mic} \mathcal{M}(Y_2, \lambda)|_{\Lambda'}$. But this is impossible since these modules are inequivalent. Indeed, they have different supports.

□

<u>Step 3.</u> The hypothesis "$\mathbf{n} \cap \mathbf{s} \cap C_p$ is non-empty" is not satisfied for certain discrete series representations on symmetric spaces, and hence theorem 4.6 does not apply directly. However, we can reduce the problem to theorem 4.6 using induction by stages. This method was used by [Vogan 5] to prove theorem 4.1 above. We will follow the same approach.

First, we need an effective way to check when is $\mathbf{n} \cap \mathbf{s} \cap C_p$ non-empty. Using the notation of section 3, if α is a root of the adjoint action of $\mathbf{a}$ on $\mathbf{g}$, we denote by $\mathbf{g}_\alpha$ the corresponding root space. $\mathbf{g}_\alpha$ is usually not one dimensional; it is in fact a representation of $\ell = \mathbf{m} + \mathbf{a}$. An element z of $\mathbf{g}_\alpha$ is called *generic* if $[z, \mathbf{g}_\alpha]$ contains a non-zero element of $\mathbf{a}$. Let $\alpha_1, \ldots, \alpha_r$ be a set of simple roots for the action of $\mathbf{a}$ in $\mathbf{n}$; pick in each root space $\mathbf{g}_{\alpha_i}$ a generic element z_i, for $i = 1, \ldots, r$. [Kostant-Rallis] have proved that the element $z = \sum_{i=1}^{r} z_i$ belongs to $\mathbf{n} \cap C_p$.

4.7 Lemma: [Vogan 5, Lemma 6.10] *If* $\mathbf{g}_\alpha$ *is not compact (i.e.* $\mathbf{g}_\alpha \cap \mathbf{s} \neq 0$*), then* $\mathbf{g}_\alpha \cap \mathbf{s}$ *contains a generic element.*

Proof: Let $\alpha^\vee$ be the coroot in $\mathbf{a}$ corresponding to α. Let $< .,. >$ denote the Killing form of $\mathbf{g}$. Recall that the involution θ acts trivially on $\mathbf{a} \subset \mathbf{k}$, while σ acts on $\mathbf{a}$ by $-id$. Define a non-degenerate symmetric bilinear form B on $\mathbf{g}_\alpha$ by

$$B(y, z) = < \alpha^\vee, [y, \sigma z] > .$$

Since θ preserves the form B, the restriction of B to each eigenspace of θ is non-degenerate. By hypothesis, the (-1)-eigenspace is non-zero. Therefore, we can choose $z \in \mathbf{g}_\alpha \cap \mathbf{s}$ such that $B(z, z)$ is non-zero, and hence $[z, \sigma z] \neq 0$. On the other hand, $[z, \sigma z]$ belongs to the (-1)-eigenspace of σ in ℓ, which is $\mathbf{a}$. Therefore, z is generic. □

This lemma yields the following dichotomy, see [Vogan 5, prop. 6.11].

1. If some of the root spaces $\mathbf{g}_\alpha$ are compact, then there exists a θ-stable proper parabolic $\mathbf{p}' = \ell' + \overline{\mathbf{n}'}$ properly containing $\mathbf{p}$, such that ℓ' is σ-stable, and L'_o/L_o is compact, where $L'_o = L' \cap G_o$, etc. In other words, if Y denotes the closed K-orbit of $\mathbf{p}$ in X, there is a "smaller" flag variety X' containing the closed K-orbit Y' of $\mathbf{p}'$, and Y' is the image of Y by the natural surjection $\gamma : X \to X'$. Put $\rho(\ell'/\ell) := \rho_\ell - \rho_{\ell'}$. This shift appears because of our choice of normalizations. Since L'_o/L_o is compact, it is clear from the Borel-Weil theorem, that $\gamma_* \mathcal{M}(Y, \lambda)$ is zero as a $(\gamma_* \mathcal{D}_\lambda, K)$-module, unless $\lambda + \rho(\ell'/\ell)$ is dominant, regular and integral as an infinitesimal character for ℓ'. Now, observe that we can write the global section functor as a composition of direct image functors:

$$\Gamma(X,.) = \gamma_{X,*}^{\{point\}} = \gamma_{X',*}^{\{point\}} \circ \gamma_{X,*}^{X'}$$

where $\gamma_X^{\{point\}}$ and $\gamma_{X'}^{\{point\}}$ are the projections of X and X' onto a point, respectively. It follows that the global sections of $\mathcal{M}(Y, \lambda)$ are zero, unless the $\lambda + \rho(\ell'/\ell)$ is dominant, regular and integral as an infinitesimal character for ℓ'.

2. If all root spaces $\mathbf{g}_\alpha$ are non-compact, then $\mathbf{n} \cap \mathbf{s} \cap C_p$ is not empty. Then, theorem 4.6 applies, and the $(\mathbf{g}, K)$-modules $M(Y, \lambda)$ are irreducible and mutually inequivalent for different K-orbits Y.

Thus, the study of the irreducibility and inequivalence of the $(\mathbf{g}, K)$-modules $M(Y, \lambda)$ can be performed inductively on the dimension of $\mathbf{g}$. One manner of finishing the proof of theorems 4.1 and 4.2, is to continue as in [Vogan 5], by listing the weights λ that need to be examined, for each simple Lie group. It turns out that this inductive procedure takes care of all cases associated to the exceptional groups. For the remaining classical groups, namely the symplectic groups, one has to invoke further arguments based on the translation principle.

□

Another way to complete the proof of theorems 4.1 and 4.2 is to use twisted sheaves of matrix differential operators. Since we only sketched their definitions in chapter I, this argument can be considered only as a sketch of proof. All their properties needed here can be proved in the same way as for twisted sheaves of differential operators. Suppose we are in the first case of the dichotomy.

Consider the diagram

$$\begin{array}{ccc} Y & \xrightarrow{i} & X \\ \delta \downarrow & & \downarrow \gamma \\ Y' & \xrightarrow{j} & X'. \end{array}$$

The fibers of δ and γ are both isomorphic to $F := L'/P \cap L$. The crucial point is that L'_o/L_o is compact, therefore F is also equal to $(L' \cap K)/(P \cap K)$, and $\gamma^{-1}(Y') = Y$. The $(\mathcal{D}, K)$-module $\mathcal{M}(Y, \lambda)$ is by construction $i_* \mathcal{O}_Y(\lambda)$ for some K-homogeneous line bundle $\mathcal{O}_Y(\lambda)$ on Y, see section II.7. By commutativity of the above diagram, we have:

$$\gamma_* i_* \mathcal{O}_Y(\lambda) = j_* \delta_* \mathcal{O}_Y(\lambda).$$

Suppose that $\lambda + \rho(\mathfrak{l}'/\mathfrak{l}) \in \mathbf{t}^*$ is dominant regular integral as a weight for L'. In the contrary case, we know that the representation $M(Y, \lambda)$ is zero, and there is nothing to prove. Let V be the finite dimensional representation of $L' \cap K$ with infinitesimal character $\lambda + \rho(\mathfrak{l}'/\mathfrak{l})$. The sheaf of rings $\delta_* \mathcal{D}_\lambda$ is isomorphic to the sheaf of differential operators acting on the local sections of the homogeneous vector bundle over Y' with fiber V. It is locally generated by the action of the envelopping algebra of $\mathbf{k}$. The Borel-Weil theory implies that the sheaf $\delta_* \mathcal{O}_Y(\lambda)$ is an irreducible sheaf of modules over this sheaf of rings.

The advantage of working with X' and Y' is that the moment map from X' to $\mathbf{g}$ has better properties than the moment map of X. Namely, the intersection $\mathbf{n}' \cap \mathbf{s} \cap C_{p'}$ is closer to be non-empty than before, where $\mathbf{p}'$ is the base point of Y', and $\mathbf{n}'$ is its opposite nilpotent radical. Let us denote by $\mathbf{a}'$ the (-1)-eigenspace of σ in the center of $\mathfrak{l}'$. Then, the sum of the dimensions of the compact root spaces of $\mathbf{a}'$ in $\mathbf{g}$ is smaller than the previous sum of dimensions for $\mathbf{a}$. Continuing in this way to eliminate the compact root spaces of $\mathbf{a}'$, we eventually reach the second case of the dichotomy. Now, we can generalize theorem 4.6 to the case of the matrix differential operators considered here. This is not difficult, although it merits to be written down explicitely. We conclude that the $(\mathbf{g}, K)$-moodules $M(Y, \lambda)$ are irreducible and inequivalent for different K-orbits Y.

□

One can also prove theorem 4.6 using the following algebraic result which extends proposition 6.5 in [Vogan 4]. We prove it here for the convenience of the reader. Recall that a primitive ring is a ring whose zero ideal is primitive, *i.e.* the ring possesses a faithful simple module.

4.8. Proposition: *If* $R_1 \subset R_2$ *are two primitive rings such that the left and right annihilators of* R_2/R_1 *in* R_1 *are non-zero, then* R_1 *and* R_2 *have the same faithful simple modules.*

Proof: Let J be the right annihilator of R_2/R_1 in R_1. All tensor products will be defined over R_1. Suppose M is a faithful simple R_2-module. Consider a faithful simple R_1-submodule L of M. Then $JL = L$ and hence $R_2/R_1 \otimes L = 0$. By right exactness of the tensor product, we obtain a surjection $L = R_1 \otimes L \twoheadrightarrow R_2 \otimes L = M$. Thus $L = M$ and M is a simple R_1-module.

Now suppose M is a faithful simple R_1-module. Then $JM = M$ and hence we obtain a surjection as above $f : M = R_1 \otimes M \twoheadrightarrow R_2 \otimes M =: N$. Since M is simple, N is either isomorphic to M or zero as R_1-module. Now $K := \ker f$ is a quotient of $\mathrm{Tor}(R_2/R_1, M)$. As an R_1-module, K is annihilated by the left annihilator of R_2/R_1, which is not zero. Since M is faithful, K is not equal to M. Thus f is non-zero and $M = N$, that is every faithful simple R_1-module has a unique structure of R_2-module.

Finally, suppose M_1, M_2 are two faithful simple R_2-modules. We have shown that each M_i remains simple as an R_1-module. Let $f_1 : M_1 \to M_2$ be an R_1-isomorphism. Then f_1 induces an R_2-map $f_2 : R_2 \otimes M_1 \to R_2 \otimes M_2$ which is non-zero since f_2 restricted to $1 \otimes M_1$ is f_1. We proved above that $M_i = R_2 \otimes M_i$. Thus f_2 is an R_2-isomorphism and $M_1 \simeq M_2$. □

The hypotheses are verified as follows. Let $R_2 = D_\lambda$ and $R_1 = U_{\chi_\lambda} = op(U)$ in the notation of III.6. Since the moment map is birational, it follows from [Borho-Brylinski I, corollary 5.12] that R_2/R_1 has an associated variety strictly smaller than $\mathcal{N}_p$: the image of the moment map. Hence, the left and right annihilators of R_2/R_1 in R_1 are non-zero. Finally, $M(Y, \lambda)$ is a faithful module for R_1 and R_2 if and only if $\mathbf{n} \cap \mathbf{s} \cap C_p$ is non-empty.

IV.5 Intertwining maps between distinct eigenspaces of $\mathbf{D}(G/H)$

Let us recall the problem which motivated our work. If G_o/H_o is a connected reductive real symmetric space, then we would like to prove that the discrete spectrum of $L^2(G_o/H_o)$ is a multiplicity free representation of G_o. In the previous section, we obtained corollary 4.3 which readily implies that the if we consider the commuting actions of both G_o and $\mathbf{D}(G/H)$ on the discrete part of $L^2(G_o/H_o)$, then it is indeed a multiplicity one representation. However, it would be nicer to be able to distinguish the discrete components of $L^2(G_o/H_o)$ solely by the action of G_o. We can do this for classical groups and a few others as stated in the following theorem.

5.1 Theorem: *Let G_o/H_o be a connected reductive real symmetric space which does not contain any factors of type $E_6/\mathbf{so}(10)\oplus\mathbf{C}$, E_6/F_4, $E_7/E_6\oplus\mathbf{C}$, or $E_8/E_7\oplus\mathbf{sl}(2)$. Then the discrete spectrum of $L^2(G_o/H_o)$ is a multiplicity free representation of G_o.*

Indeed, proposition 3.3 implies that under the hypothesis considered here, the eigenspaces of $\mathbf{D}(G/H)$ have different infinitesimal characters for $Z(\mathbf{g})$. Then, it suffices to aplly corollary IV.4.3. By the same reasoning, we also obtain:

5.1' Theorem: *Let G_o/H_o be a connected reductive symmetric space which is of type $E_6/\mathbf{so}(10)\oplus\mathbf{C}$, E_6/F_4, $E_7/E_6\oplus\mathbf{C}$, or $E_8/E_7\oplus\mathbf{sl}(2)$. Then the discrete spectrum of $L^2(G_o/H_o)$ is a multiplicity free representation of G_o, provided we only consider representations with infinitesimal characters which are not ambiguous in the sense of section 3. Moreover, for any ambiguous infinitesimal characters χ, the discrete spectrum part of the two distinct eigenspaces of $\mathbf{D}(G/H)$ with infinitesimal character χ, each are separately multiplicity free.*

If a representation V has a non-ambiguous infinitesimal character χ, then χ determines exactly in which eigenspace of $\mathbf{D}(G/H)$ in $C^\infty(G_o/H_o)$ or $L^2(G_o/H_o)$, V can be imbedded. Otherwise, there are two possible eigenspaces, and we have not ruled out the possibility that V admit an imbedding in both eigenspaces. Recall that we have listed the ambiguous infitesimal characters in the proof of lemma 3.5, and they occur only for the exceptional symmetric spaces of theorem 5.1'.

The connectedness hypothesis is important since there are examples of symmetric quotients for finite groups having multiplicities bigger than one, *e.g.* $G = SL_2(\mathbf{F}_3)$ and $H=$ diagonal matrices, then $G/H = PSL_2(\mathbf{F}_3)$ which affords a representation with multiplicity three.

Concerning the general case of the problem mentioned at the beginning of this section, we shall shall prove the following result which has some interest of its own. Its proof was found in collaboration with Joseph Bernstein. Before stating the result, we have to describe a technical hypothesis that will be needed for the proof. The description of this hypothesis will also introduce some of our notation.

Put $S_o := G_o/H_o$. By definition, an endomorphism of a topological vector space is continous. Now, by restriction, an endomorphism ℓ of $L^2(S_o)$ maps continuously the subspace $C_c^\infty(S_o)$ of smooth functions with compact support on S_o, into the space $C^{-\infty}(S_o)$ of distributions on S_o, which contains $L^2(S_o)$. Recall L. Schwartz's kernel theorem, see [Treves, p. 531]. It asserts that any (continuous) morphism ℓ from $C_c^\infty(S_o)$ into $C^{-\infty}(S_o)$ is given by integration with a distribution kernel L on $S_o \times S_o$. Although L is originally defined on the real variety $S_o \times S_o$, we can extend it to the complex variety $S \times S$, where $S = G/H$. Consider as before a maximal σ-split torus A in G. Let A_{reg} denote the regular elements of A, *i.e.* the elements of A on which the roots of A in $\mathbf{g}$ do not take the value 1. The subset $G_{reg} := HA_{reg}H$ is open in G, and so is the subset $S_{reg} := (HA_{reg}H)/H$ in S. Moreover, the space $H\backslash S_{reg} := H\backslash G_{reg}/H$ is equal to $W_a\backslash A_{reg}$, where W_a is the Weyl group of A in $\mathbf{g}$. Recall that for a group G and a subgroup H, there is a bijection between the double coset spaces:

$$G\backslash(G/H \times G/H) \simeq H\backslash G/H .$$

It is defined as follows: any G-orbit on $G/H \times G/H$ contains a point of the form (e,x) where e is the identity element of G. Then, the G-orbit of (e,x) corresponds to the H-orbit of x in G/H. Thus, there is an open subset $(S \times S)_{reg}$ of $S \times S$ whose quotient by the diagonal action of G is isomorphic to $W_a \backslash A_{reg}$. The singular set of $S \times S$ is by definition the complement of the regular set $(S \times S)_{reg}$. For semisimple symmetric spaces, the codimension of the singular set is $1 + \dim G/P \geq 2$, where P is a parabolic subgroup associated to H by an Iwasawa decomposition of $\mathbf{g}$. We say that the kernel L is *regular* if it is determined by its restriction to the regular set $(S \times S)_{reg}$. Our technical hypothesis is to work with regular kernels.

5.2 Theorem: *If ℓ is an endomorphism of $L^2(G_o/H_o)$ commuting with the group G_o, and whose kernel is regular, then ℓ also commutes with the algebra* $\mathbf{D}(G/H)$.

Before proving this theorem, let us mention that if we could get rid of the technical hypothesis, then this result would give the solution to our problem: namely, that the discrete series of all connected reductive symmetric spaces is multiplicity free. Indeed, the eigenspaces corresponding to the discrete spectrum of $\mathbf{D}(G/H)$ are direct summands in $L^2(G_o/H_o)$. Moreover they are themselves finite direct sums of their irreducible constituents. We already know by corollary 4.3 that within one such eigenspace, all constituents are inequivalent. If an irreducible representation V of G_o would appear in two different eigenspaces, then composing one inclusion of V into $L^2(G_o/H_o)$ with the projection of $L^2(G_o/H_o)$ onto the other copy of V, we would obtain a non-zero endomorphism ℓ of $L^2(G_o/H_o)$ commuting with G_o. According to 5.2, ℓ must also commute with $\mathbf{D}(G/H)$. But ℓ does not, since it changes the eigenvalues of the elements of $\mathbf{D}(G/H)$. Thus, V has multiplicity one in the discrete spectrum of $L^2(G_o/H_o)$.

There exist examples of H_o-invariant eigendistributions on G_o/H_o which are supported on the singular set, *cf.* [Kengmana] where such examples are announced for $SO(p,q+1)/SO(p,q)$, with p, q even. These distributions could produce interwining maps which are not regular. Therefore, theorem 5.2 as stated does not solve the full generality of our problem. Nevertheless, its proof reveals several interesting features of the matter.

Proof: The symmetric subgroup H acts freely on both sides of the group G. Put $S = G/H$, so that H still acts on the left of S, but no longer freely. Consider the affine algebraic variety $S \times S$. It has a left action of G on the first factor – denoted by α_1, and it also has a left action of G on the second factor – denoted by α_2. Let α denote the diagonal action of G on $S \times S$. Consider the sheaf of rings $\mathcal{D}_{S\times S}$ of differential operators on $S \times S$. Then, there is a morphism $d\alpha : \mathbf{g} \to \Gamma(S \times S, \mathcal{D}_{G\times G})$. In I.2.2, we have defined the category of $(\mathcal{D}_{S\times S}, G)$-modules.

Let $\mathcal{F}$ be a $(\mathcal{D}_{S\times S}, G)$-module. The algebra $\mathbf{D}(S) = \mathbf{D}(G/H) = U(\mathbf{g})^H / U(\mathbf{g})\mathbf{h} \cap U(\mathbf{g})^H$ of G-invariant differential operators on S is a subalgebra of $\Gamma(S, \mathcal{D}_S)$; hence it acts on any $\mathcal{D}_S$-module. Therefore, $\mathcal{F}$ has a left action α_1 , and a right action α_2 of $\mathbf{D}(S)$. There is a natural pairing on $C_c^\infty(G_o/H_o)$ which induces the Hilbert space structure on $L^2(G_o/H_o)$. One can define the adjoint A^* of an operator $A \in End(L^2(G_o/H_o))$ with respect to this pairing. Similarly, there is a natural anti-involution $*$ on $U(\mathbf{g})$ which is $-Id$ on $\mathbf{g}$. These two operations are consistent, so that if $d \in \mathbf{D}(S)$, then we have: $\alpha_i(d)^* = \alpha_i(d^*)$, for $i = 1,\ 2$. Let $\Gamma\mathcal{F}^G$ denote the sections of $\mathcal{F}$ on $S \times S$ which are invariant by $\alpha(G)$. This is again a $\mathbf{D}(S)$-module.

5.3 Proposition: *Let* $v \in \Gamma\mathcal{F}^G$, *and* $d \in \mathbf{D}(S)$. *Then, the following equality holds over the regular set* $(S \times S)_{reg}$:

$$\alpha_1(d)v = \alpha_2(d^*)v \ .$$

Proof: There exists a unique $\mathcal{D}_{S\times S}$-module $\mathcal{R}$ generated by a single G-fixed section – denoted r. Explicitly, $\mathcal{R}$ is defined by:

$$\mathcal{R} = \mathcal{D}_{S\times S}/\mathcal{I}$$

where $\mathcal{I}$ is the two-sided ideal given by $\mathcal{D}_{S\times S} \cdot \mathbf{g}$. The generating section r is the constant section $1 + \mathcal{D}_{S\times S} \cdot \mathbf{g}$.

Let $i : (S \times S)_{reg} \hookrightarrow S \times S$ be the inclusion. Consider first the restriction $\mathcal{R}|_{(S\times S)_{reg}}$ of $\mathcal{R}$ to the regular set, and then the direct image module $\mathcal{R}_{reg} := i_*(\mathcal{R}|_{(S\times S)_{reg}})$. There is a natural map $\mathcal{R} \to \mathcal{R}_{reg}$ which is defined as follows. $\mathcal{R}$ is generated over $\mathcal{D}_{S\times S}$ by the constant section $r = 1 + \mathcal{I}$. Since $i : (S \times S)_{reg} \to S \times S$ is an open imbedding, $\mathcal{R}$ and $\mathcal{R}_{reg}$ coincide over the open set $(S \times S)_0$. Therefore, $\mathcal{R}_{reg}$ also contains the section r that generates R over $\mathcal{D}_{S\times S}$. To avoid confusion, in later reference, let us denote by r_o, the section r seen as a section of $\mathcal{R}_{reg}$. By applying $\mathcal{D}_{S\times S}$ to r_o inside $\mathcal{R}_{reg}$, we obtain the map announced. The kernel of this map is given by the largest submodule of $\mathcal{R}$ whose sections are supported on the singular set $Y := S \times S \setminus (S \times S)_0$.

Now, the space of G-invariant sections $\mathcal{R}_{reg}$ forms a regular $\mathbf{D}(S)$-module, since $i_*(\mathcal{R}|_{(S\times S)_0})$ does not have any section supported on the singular set. We can easily determine this space. From the definitions of $\mathcal{R}$ and of the direct image functor i_*, we find:

$$\Gamma\mathcal{R}^G_{reg} := \Gamma i_*(\mathcal{R}|_{(S\times S)_0})^G = \Gamma(\mathcal{R}|_{(S\times S)_{reg}})^G \simeq D_{(W_a\backslash A_{reg})} = \Gamma\mathcal{O}_{(W_a\backslash A_{reg})} \otimes S(\mathbf{a})^{W_a} \, .$$

where $D_{(W_a\backslash A_{reg})}$ denotes the algebra of all polynomial differential operators on $D_{(W_a\backslash A_{reg})}$. Since $\mathbf{D}(S) \simeq S(\mathbf{a})^{W_a}$ by Harish-Chandra's isomorphism, it follows that $\Gamma(\mathcal{R}|_{(S\times S)_{reg}})^G$ is a free $\mathbf{D}(S)$-module. In particular, the $\mathbf{D}(S)$-submodules of $\Gamma\mathcal{R}^G_{reg}$ generated by the section r_o with the the actions α_1 and α_2 are both free.

We proceed to the proof of the proposition for $\mathcal{F} = \mathcal{R}_{reg}$. The bi-invariant differential operators on G form the ring $Z(\mathbf{g})$ which maps into $\mathbf{D}(S)$; let $Z(S)$ be its image. The assertion 5.3 is clear if d belongs to $Z(S)$. Indeed, we can let these operators act on functions on S either from the right or the left. But if the left translation is an action, then the right translation obtained by commutation of the operator with the group elements, is an anti-action. This has the effect of changing the left action of an element $d \in Z(S)$ into the right action of its adjoint, *cf.* [Helgason 3, II.5.31]. Moreover the G-invariance for an invariant section $v \in \Gamma\mathcal{R}^G_{reg}$ means that for all $g \in G$, $x, y \in S$, we have:

$$v(g \cdot x, y) = v(x, g^{-1} \cdot y) \, .$$

The proposition follows for all elements of $Z(S)$.

Recall Helgason's result 3.3: for any element $d \in \mathbf{D}(S)$, there exists two elements z_1 and z_2 in $Z(S)$, such that $d \cdot z_1 = z_2$ as element of $\mathrm{End}(\Gamma\mathcal{R}^G_{reg})$. We have:

$$\alpha_1(d \cdot z_1) - \alpha_2((d \cdot z_1)^*) = \alpha_1(d)\alpha_1(z_1) - \alpha_2(z_1^*)\alpha_2(d^*)\,.$$

Using the commutativity of $\mathbf{D}(S)$, and the fact that the proposition is true for the elements of $Z(S)$, we obtain:

$$\alpha_1(z_1)(\alpha_1(d) - \alpha_2(d^*))\,.$$

But this must be equal to:

$$\alpha_1(z_2) - \alpha_2(z_2^*) = 0$$

Since $\Gamma\mathcal{R}^G_{reg}$ is a free $\mathbf{D}(S)$-module, it does not contain any torsion. Therefore by applying $\alpha_1(z_1)(\alpha_1(d)-\alpha_2(d^*))$ to r_o, we conclude that $\alpha_1(d)r_o = \alpha_2(d^*)r_o$.

The following fact shows the importance of the module $\mathcal{R}$, and its generating section r.

5.4 Lemma: *Let $\mathcal{F}$ be a $(\mathcal{D}_{S\times S}\,,\, G)$-module. Then:*

$$\Gamma\mathcal{F}^G = \mathrm{Hom}_{(\mathcal{D}_{S\times S}\,,G)}(\mathcal{R}, \mathcal{F})\,.$$

Proof: This is a kind of Fröbenius reciprocity. If $v \in \Gamma\mathcal{F}^G$, then the corresponding homomorphism ϕ_v maps r to v, and is determined by this property. Conversely if v is the image of r by a homomorphism ϕ, then $v \in \Gamma\mathcal{F}^G$. The last statement is proved in the same manner. □

This lemma implies proposition 5.3. Indeed, for $v \in \Gamma\mathcal{F}^G$, let ϕ be the homomorphism given in the proof of lemma 5.4, such that $v = \phi(r)$. Since v is a regular section in the module $\Gamma\mathcal{F}^G$, it is determined by its restriction v_o the regular set $(S \times S)_{reg}$. We can also localize ϕ obtain a homomorphism ϕ_o over the regular set such that $v_o = \phi_o(r_o)$. For any $d \in \mathbf{D}(S)$, we have:

$$\alpha_1(d)v_o = \alpha_1(d)\phi_o(r_o) = \phi_o(\alpha_1(d)\,r_o)$$

$$= \phi_o(\alpha_2(d^*)\,r_o) = \alpha_2(d^*)\phi_o(r_o) = \alpha_2(d^*)v_o\,,$$

using the fact that ϕ_o commutes with differential operators on $(S \times S)_{reg}$. □

To finish the proof of theorem 5.2, we use L. Schwartz's kernel theorem to reprroent the intertwining map ℓ by a distribution kernel on $S_o \times S_o$. Since the singular set has codimension at least two, it has measure zero. The regularity assumption on L implies that we can write ℓ as an integral over the regular set. For $y \in S_o$, define $S_{o,reg}(y) := \{(y,x) \in (S_o \times S_o)_{reg}\}$. Given $f \in C_c^{infty}(S_o)$, we have:

$$\ell f(y) = \int_{S_{o,reg}(y)} L(y,x) f(x) dx,$$

where dx is a G_o-invariant measure on S_o. Since ℓ commutes with the left action of G_o, L is invariant under the diagonal action of G_o. Although L is originally defined on the real variety $S_o \times S_o$, we can extend this distribution to $S \times S$, and apply to it all operators in $\mathcal{D}_{S\times S}$. Let $\mathcal{F} := \mathcal{D}_{S\times S} \cdot L$ be the $\mathcal{D}_{S\times S}, G)$-module generated by L, so that $L \in \Gamma\mathcal{F}^G$. Thanks to proposition 5.3, the following equality holds over the regular set for all $d \in \mathbf{D}(S)$:

$$\alpha_1(d)L \;=\; \alpha_2(d^*)L \;.$$

For a function $f \in C_c^\infty(S_o)$, and an invariant differential operator $d \in \mathbf{D}(S)$, denote by $d \cdot f$ the natural action of d on f. Then we have:

$$\begin{aligned} \ell(d \cdot f)(y) &= \textstyle\int_{S_{o,reg}(y)} L(y,x)\; d \cdot f(x)\; dx \\ &= \textstyle\int_{S_{o,reg}(y)} \alpha_2(d^*)L(y,x)\; f(x)\; dx \\ &= \textstyle\int_{S_{o,reg}(y)} \alpha_1(d)L(y,x)\; f(x)\; dx \\ &= d \cdot (\ell f)(y)\,. \end{aligned}$$

Since $C_c^\infty(S_o)$ is dense in $L^2(S_o)$, it follows by continuity of ℓ, that ℓ commutes with every element of $\mathbf{D}(S)$ on the whole space.

□

5.5 Remark: We can generalize theorem 5.2 to the C^∞ context in the following way . More precisely, let ψ be a character of $\mathbf{D}(G/H)$, and define

$$C^\infty(G_o/H_o;\psi) := \{f \in C^\infty(G_o/H_o) |\ \nu(d)\cdot f = \psi(d)\, f \,, \text{for all } d \in \mathbf{D}(G/H)\}$$

be the ψ-proper-subspace of $\mathbf{D}(G/H)$ acting on $C^\infty(G_o/H_o)$, and let

$$C^\infty(G_o/H_o;\overline{\psi}) := C^\infty(G_o/H_o)/\{\nu(d)\cdot f - \psi(d)\, f |\ f \in C^\infty(G_o/H_o)\,,\ d \in \mathbf{D}(G/H)\}$$

be the ψ-proper-quotient-space of $\mathbf{D}(G/H)$ acting on $C^\infty(G_o/H_o)$. Then we obtain for two different characters $\psi_1 \neq \psi_2$ of $\mathbf{D}(G/H)$:

$$\mathrm{Hom}_{G_o}(C^\infty(G_o/H_o;\psi_1)\,,\ C^\infty(G_o/H_o;\overline{\psi_2})) = 0\,.$$

We could allow generalized eigenspaces, provided we fix the degree of nilpotence of $\mathbf{D}(G/H)$. This does not imply that between different eigenspaces of $\mathbf{D}(S)$, there cannot be any interwining map. Indeed, a proper subspace need not isomorphic to the corresponding proper quotient-space. For example, the reader can contemplate the case of the zero eigenvalue of the operator $\frac{d}{dx}$ acting on the space $C^\infty(\mathbf{R})$. Then:

$$C^\infty(\mathbf{R};0) = \mathbf{R}, \quad \text{while} \quad C^\infty(\mathbf{R};\overline{0}) = 0.$$

5.6 Scholium: From the proof of theorem 5.2, we see clearly that the regularity hypothesis on the kernel of ℓ could be waived, if we could prove the following statement. Let A_i be the submodule generated by the α_i-action of $\mathbf{D}(S)$ on r in $\Gamma\mathcal{R}^G$, for $i = 1, 2$. *The sum of* A_1 *and* A_2 *in* $\Gamma\mathcal{R}^G$ *is a free* $\mathcal{D}(S)$*-module for each action* α_1 *and* α_2.

We have not been able to prove this assertion, and we offer it to the reader as a challenge. A stronger statement that may hold is that $\Gamma\mathcal{R}^G$ itself is a free $\mathcal{D}(S)$-module for each action α_1 and α_2. If this assertion is true, it is likely to be a special feature of semisimple symmetric spaces, somewhat related to the fact that although K does not act freely on G/K, the ring of regular functions on G/K is free over the subring of K-invariant functions, *cf.* [Kostant-Rallis].

IV.6 Multiplicities of principal series

We do not assume any longer that $\operatorname{rank} G/H = \operatorname{rank} K/H \cap K$. Among the other subrepresentations which occur in $A_K(S_o; \psi_\lambda)$ are those which belong to the principal series. They correspond to the $(\mathbf{g}, K^\tau)$-submodules of $B_{K^\tau}(X^\tau; \mathbf{C})$ which are supported by the open K^τ-orbit X^τ or rather by their closures. Let Y^τ be an open K^τ-orbit different from the whole variety X^τ and let $B_{Y^\tau}(X^\tau; \mathbf{C})$ denote the subspace of K^τ-algebraic hypersections which are supported in the closure of Y^τ. We say that λ is generic if it is non-integral in $\mathbf{a}^*$ except on the roots of $\mathbf{a}$ in $\mathbf{k}$ and if $\lambda + \rho_\ell$ is dominant regular in $\mathbf{t}^*$. Then $B_{Y^\tau}(X^\tau; \mathbf{C}_\lambda)$ is an irreducible $(\mathbf{g}, K^\tau)$-module and hence it corresponds to an irreducible subrepresentation $V(Y, \lambda)$ of G_o in $A_K(G_o/H_o; \psi_\lambda)$ via Helgason's and Flensted-Jensen isomorphisms. On the other hand, let Y_0 be the unique open K-orbit in X. Then

$$Y_0 \cap X^\tau = Y_1^\tau \cup \cdots \cup Y_n^\tau$$

is a disjoint union of say n open K^τ-orbits in X^τ.

6.1 Theorem: *If λ is generic, $V(Y, \lambda)$ occurs in $A_K(G_o/H_o; \psi_\lambda)$ with multiplicity n.*

A similar result holds for every K-orbit Y in X whose intersection with X^τ is non-empty. In particular, if Y is closed and intersects the open H-orbit X^0, then $Y \cap X^\tau$ consists of exactly one K^τ-orbit, cf. II.7.4. This explains why the discrete series of G_o/H_o is multiplicity free.

Proof: Consider the inclusion $I : Y_0 \to Y$. When λ is generic the two direct image functors $i_!$ and i_* coincide. The stabilizer K_y of a point $y \in Y_0$ may be disconnected, and to construct a $(\mathcal{D}_\lambda, K)$-module $\mathcal{M}(Y_0, \lambda)$ as in §3 we must choose a representation of M which is trivial on its identity component and on $H \cap K_y$. Then the $(\mathcal{D}_\lambda, K)$-module $\mathcal{M}(Y_0, \lambda)$ is irreducible. Since $\lambda + \rho_\ell$ is dominant regular in $\mathbf{t}^*$, the global sections of $\mathcal{M}(Y_0, \lambda)$ form an irreducible $(\mathbf{g}, K)$-module denoted $M(Y_0, \lambda)$.

Now we can refine theorem II.6.8 to obtain an injective map of $(\mathbf{g}, K^r)$-modules

$$R : M(Y_0, \lambda) \longrightarrow B_{K^r}(X^r; \mathbf{C}) .$$

Indeed, X^r is contained in the open H-orbit X^0. On X^0 there is an invertible sheaf $\mathcal{O}_{X^0}(\lambda)$ whose local cohomology along X^r gives the space of hypersections $B(X^r; \mathbf{C}_\lambda)$. G^r acts on this space and $B_{K^r}(X^r; \mathbf{C}_\lambda)$ consists of the K^r-algebraic elements. On the other hand, since the global sections of $M(Y_0, \lambda)$ are K-algebraic, they are determined by their restriction to $Y_0 \cap X^r$. Then they are mapped into $B_{K^r}(X^r; \mathbf{C}_\lambda)$ by the same process as in II.6. The image of R consists of vectors which are supported on all of $Y_0 \cap X^r$ because $M(Y_0, \lambda)$ is irreducible and hence it has no constituent concentrated on strictly smaller K-orbits than Y_0.

Similarly, the genericity of λ forces the hypersections in $B_{K^r}(X^r; \mathbf{C}_\lambda)$ to have singularities along the boundary of Y_0 and prevents the occurrence of hypersections supported on strictly smaller K-orbits. Therefore, multiplication by the characteristic function of the closure of any open K^r-orbit Y_i^r, $i = 1, \ldots, n$, is an operator which commutes with the actions of $\mathbf{g}$ and K^r. In this way, we obtain n projections of $M(Y_0, \lambda)$ on the various submodules $B_{Y^r}(X^r; \mathbf{C}_\lambda)$ $i = 1, \ldots, n$. These projections must be bijective because acting with the complex group K on any $B_{Y_i^r}(X_i^r; \mathbf{C}_\lambda)$ gives back the original module $M(Y_0, \lambda)$. Thus we have shown that all the $(\mathbf{g}, K^r)$-modules $B_{Y_i^r}(X^r; \mathbf{C}_\lambda)$ are isomorphic. Consequently, $V(Y, \lambda)$ occurs with multiplicity n in $A_K(G_o/H_o; \psi_\lambda)$. □

It is clear from the result that the appearance of multiplicity for symmetric spaces is due to the fact that the real points of the orbit of the complex group K split into several orbits for the corresponding real form of K. This kind of situation is familiar in the study of conjugacy classes where it has led to the notion of stable conjugacy. In fact, one can parametrize the representations in $L^2(G_o/H_o)$ by coadjoints orbits of H_o in $\mathrm{Lie}(H_o)^\perp \subset \mathrm{Lie}(G_o)^*$. Now, take a complex H-orbit Y in $\mathbf{h}^\perp$ seen as a subspace $\mathbf{g}$. Y intersected with the real subspace $\mathrm{Lie}(H_o)^\perp$ may split into several orbits for the real group H_o. This yields an alternative geometric explanation for multiplicities of representations in $L^2(G_o/H_o)$. However, the reader should be warned that this explanation is only valid in the generic range of $\lambda \in \mathbf{a}^*$. For special values of the parameter λ, some jumps in the multiplicities may occur.

Finally, let us mention that a recent article of [van den Ban 2] in which he obtains an explicit construction of H-invariant distributions on principal series modules. Taking into account the fact that we work with the full centralizer H of the involution σ instead of just its identity component, his results coincides with ours on the multiplicity of principal series representations on symmetric spaces.

IV.7 τ-invariant of discrete series on exceptional symmetric spaces

In this final section, we are going to describe the τ-invariant of certain representations which occur in the discrete series of exceptional symmetric spaces. Moreover these representations have ambiguous eigencharacters for the algebra $\mathbf{D}(G/H)$, in the sense that two such representations may have different eigencharacters for $\mathbf{D}(G/H)$, but have same infinitesimal character for $Z(\mathbf{g})$, see lemma 3.5.

According to Helgason's result 3.3, there are four cases where ambiguous weights may - and in fact do - occur:

$$E_6/\mathbf{so}(10) \oplus \mathbf{C}\,,\ E_6/F_4\,,\ E_7/E_6 \oplus \mathbf{C}\ ,\ E_8/E_7 \oplus \mathbf{sl}(2)\,.$$

We will detail the computations for the case of integral regular $\lambda \in \mathbf{a}^*$. Moreover we omit those λ which are ambiguous only with some singular weight μ. This suffices to study the discrete series of symmetric spaces according to theorem 3.4.

The ambiguous regular integral weights which are ambiguous with some other regular weight are the one written below. According to the proof of lemma 3.5, there are none for the symmetric space $E_6/\mathbf{so}(10) \oplus \mathbf{C}$.

		λ	μ
E_6/F_4		(4,1)	(1,4)
$E_7/E_6 \oplus \mathbf{C}$	$m = 3, 4, \cdots$	(4,1,m)	(1,4,m-2)
		(4,1,1)	(1,3,1)
$E_8/E_7 \oplus \mathbf{sl}(2)$	$\begin{cases} m = 3, 4, \cdots \\ n = 1, 2, \cdots \end{cases}$	(4,1,m,n)	(1,4,m-2,n)
	$n = 2, 3, \cdots$	(4,1,1,n)	(1,3,1,n-1)

We are going to compute the τ-invariant of the $(\mathbf{g}, K)$-modules $A(\lambda) := A_K(G_o/H_o; \psi_\lambda)$. For reference to coherent families and τ-invariants, see [Vogan 1, 2 Chap. 7]. First note that as λ varies, these representations belong to the same coherent family.

Indeed Flensted-Jensen duality together with Helgason isomorphism identifies $A(\lambda)$ with $B_{K^r}(X^r; \mathbf{C}_\lambda)$: the subspace of K^r-algebraic hyperfunction sections of the line on $X^r = G^r/P^d$ induced from $\mathbf{C}_\lambda$. The notation is the same as in section II.3. Now the full spaces of hyperfunction sections form obviously a coherent family for G^r as λ varies. Hence so does the subspace of K^r-algebraic elements. For an integral weight λ in $\mathbf{t}^*, \mathbf{Mod}(\lambda)$ shall denote the category of $(\mathbf{g}, K)$-modules with infinitesimal character χ_λ. Recall that to a simple root α, we can associate two functors Ψ_α and Φ_α on the category of $(\mathbf{g}, K)$-modules which are called translations to and from the α-wall. These functors are not uniquely determined because we have not specified to which point of the α-wall we are moving. We can choose any integral weight which lies on the α-wall and on no other wall. Then the composite functor $\mathcal{F}_\alpha := \Phi_\alpha \Psi_\alpha$ is well-defined and it preserves infinitesimal characters because $\mathbf{Mod}(s_\alpha \chi) = \mathbf{Mod}(\chi)$ for any integral μ in $\mathbf{t}^*$ where s_α is the reflection in the root α.

Suppose λ is big enough so that $\lambda_\rho := \lambda + \rho_\ell$ is a dominant regular in $\mathbf{t}^*$. Then there exists a unique coherent family Θ of virtual $(\mathbf{g}, K)$-modules whose member $\Theta(\lambda_\rho)$ at λ_ρ is $A(\lambda)$. The *τ-invariant* of the component M of $A(\lambda) \in \mathbf{Mod}(\lambda_\rho)$ is the set of simple roots α such that $\psi_\alpha M = 0$. In fact the π-invariant is more naturally attached to the primitive ideal $\mathrm{Ann}_U(M)$ and it is under this disguise that we will use it. The key tool of the computation is the following result of [Vogan 1, Theorem 3.2]:

Suppose α and β are two simple roots of the same length such that $\alpha \in \tau(M)$ and $\beta \in \tau(M)$. Then $\mathcal{F}_\alpha(M)$ has a unique component V_α with $\beta \in \tau(V_\alpha)$. Moreover, $\alpha \in \tau(V_\alpha)$. In addition, one knows exactly the primitive ideal associated to V_α; $\Psi_\beta \mathcal{F}_\alpha(M) = \Psi_\beta(V_\alpha)$ is irreducible and M occurs as a component of $\mathcal{F}_\beta(V_\alpha)$.

In the theory of $\mathcal{D}$-modules, the τ-invariant has the following interpretation. Let $\mathcal{B}$ denote the variety of Borel subgroups of G, and let X_α be the variety of minimal parabolic subgroup of type α. There is a fibration $p_\alpha : \mathcal{B} \to X_\alpha$ with fibers $\mathbf{P}^1$. The inverse image functor $p_\alpha^!$ maps $\mathcal{D}_\lambda$-modules on X_α to $\mathcal{D}_{\lambda+\alpha/2}$-modules on $\mathcal{B}$. A $\mathcal{D}$-modules $\mathcal{M}$ has α in its τ-invariant if it is the pull-back of a $\mathcal{D}$-module on X_α. It follows that our modules $\mathcal{M}(Y,\lambda)$ have at least the simple roots of $\mathbf{m}$ in their τ-invariants, hence the same holds for their global sections $M(Y,\lambda)$. We shall write a τ-invariant as a collection of $+$ and $-$ signs written in the same order as the Dynkin diagram where (resp. $+$) denotes a root which is (resp. is not) in the τ-invariant.

Let us consider the first case: $E_6/F_4, \lambda = (4,1)$, $\mu = (1,4)$ and put $\nu = (4,4)$. We shall view λ and μ as successive degenerations of ν. The τ-invariant of the components of $A(\nu)$ can be any of the four following possibilities:

1. $\begin{smallmatrix} & & - & & \\ - & - & - & - & - \end{smallmatrix}$ 2. $\begin{smallmatrix} & & - & & \\ + & - & - & - & + \end{smallmatrix}$ 3. $\begin{smallmatrix} & & - & & \\ + & - & - & - & - \end{smallmatrix}$ 4. $\begin{smallmatrix} & & - & & \\ - & - & - & - & + \end{smallmatrix}$

We have seen that $A(\lambda), A(\mu)$ and $A(\nu)$ belong to the same coherent family Θ. In fact, we have:

$$\Theta\begin{pmatrix} & & 1 & & \\ 1 & 1 & 1 & 1 & 1 \end{pmatrix} = A(\nu)$$

$$\Theta\begin{pmatrix} & & 1 & & \\ 1 & 1 & 1 & 1 & -2 \end{pmatrix} = A(\lambda) \qquad \text{and}$$

$$\Theta\begin{pmatrix} & & 1 & & \\ -2 & 1 & 1 & 1 & 1 \end{pmatrix} = A(\pi)\,.$$

The weights λ_ρ and μ_ρ are conjugate to $\begin{pmatrix} & & 1 & & \\ 1 & 1 & 0 & 1 & 1 \end{pmatrix}$ by $s_5\ s_6$ and $s_2\ s_1$ respectively. The situation is best described by the picture below. In the notation of section IV.3, we can write $\mathbf{t}^* = \mathbf{t}_m^* \oplus \mathbf{a}^*$, where $\mathbf{t}_m$ is a Cartan subalgebra of $\mathbf{m}$. Then, we want to pass successively from the point $1 = \nu + \rho_\ell$ to the point $6 = \lambda + \rho_\ell$.

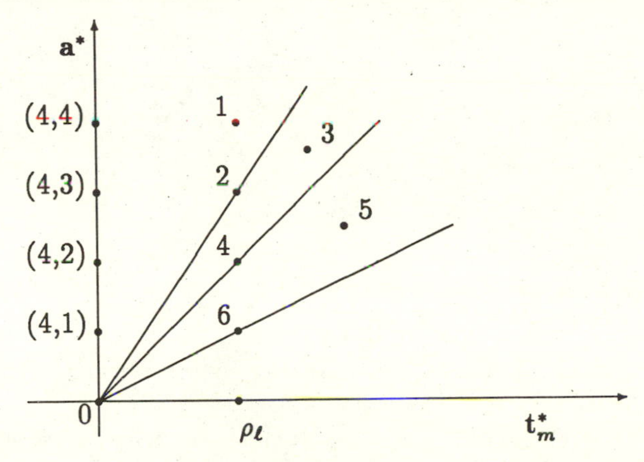

If all roots are in the τ-invariant of M, then the representations are zero at λ and μ. So let us assume that we are in the second case for τ. Recall that a constituent of a representation is an irreducible subquotient. i Let M be a constituent of $A(\nu)$. We shall work with virtual $(\mathbf{g}, K)$-modules and we shall identify them with their characters. When we push M to the α_6-wall, we get zero if $\alpha_6 \in \tau(M)$. In this case $\Theta(s_\alpha \nu) = -\Theta(\nu)$. If $\alpha_6 \tau(M)$, then $\Psi_6 M$ is a constituent of $A(4,3)$. Coming out on the other side of the α_6-wall to the weight $s_6\nu$, we see using [Vogan 1, 7.3.19], that $s_6(M) = M + U_6$ where every component of U_6 has α_6 in its τ-invariant and V_6 is the only constituent of U_6 which does not have α_5 in its τ-invariant. Thus at this stage the τ-invariants of the constituents of $s_\alpha(M)$ are:

$$\begin{aligned}
\tau(M) &= \left(\begin{smallmatrix} & & - & & \\ + & - & - & - & + \end{smallmatrix}\right), \\
\tau(U_6 - V_6) &= \left(\begin{smallmatrix} & & - & & \\ + & - & - & - & - \end{smallmatrix}\right) \quad \text{and} \\
\tau(V_6) &= \left(\begin{smallmatrix} & & - & & \\ + & - & - & + & - \end{smallmatrix}\right).
\end{aligned}$$

Now we push the representation to the wall corresponding to the fifth root of the diagram α_5. To reach the weight $\left(\begin{smallmatrix} & & 1 & & \\ 1 & 1 & 1 & 0 & 1 \end{smallmatrix}\right)$ we apply Ψ_5; then only the term $\Psi_5(V_6)$ survives. When we come out of the α_5-wall, we obtain a virtual module containing exactly one component N with τ-invariant:

$$\left(\begin{matrix} & & - & & \\ + & - & + & - & - \end{matrix}\right)$$

The τ-invariant of the rest is

$$\left(\begin{matrix} & & - & & \\ + & - & - & + & - \end{matrix}\right) \quad \text{or} \quad \left(\begin{matrix} & & - & & \\ + & - & - & - & - \end{matrix}\right)$$

α_6 is still in the τ-invariant because $s_5 s_6(\alpha_6) = \alpha_5$ belongs to the τ-invariant by hypothesis. When we apply Ψ_3, only $\Psi_3(N)$ survives with the same τ-invariant as N and at this place we obtain the weight λ_ρ.

On the other hand, if we perform the same computation starting from ν and going to μ, we get the following τ-invariant at μ_ρ:

$$\left(\begin{matrix} & & - & & \\ - & - & + & - & + \end{matrix}\right).$$

Thus starting with any constituent M of $A(\nu)$ having $\left(\begin{smallmatrix} & & - & & \\ + & - & - & - & + \end{smallmatrix}\right)$ as its τ-invariant, we obtain two different subrepresentations of $A(\lambda)$ and $A(\mu)$ respectively.

The possibilities 3 and 4 for $\tau(M)$ require a simultaneous treatment. Indeed, if $\tau(M)$ is

$$\left(\begin{matrix} & & - & & \\ + & - & - & - & - \end{matrix}\right),$$

when we go from ν to λ, the representation vanishes because it lies in walls belonging to its τ-invariant. On the other hand, going from ν to μ gives the following τ-invariant:

$$\left(\begin{matrix} & & - & & \\ - & - & + & - & - \end{matrix}\right) \tag{$*$}$$

If we take $\tau(M)$ to be $\left(\begin{smallmatrix} & & - & & \\ - & - & - & - & + \end{smallmatrix}\right)$, then going from ν to μ kills the representations, but at λ we get the same τ-invariant (*).

Two inequivalent representations with same infinitesimal character may have same τ-invariant, since there exist examples of different primitive ideals with same infinitesimal character and same τ-invariant. It is this coincidence of τ-invariants that forbids us from completing the multiplicity one theorem for $L^2(G_o/H_o)$ in 5.1'. More precisely, let $M(Y_1,\lambda)$ and $M(Y_2,\mu)$ be two $(\mathbf{g},K)$-modules of the type considered in §4, which are faithful and simple over D_λ and D_μ respectively. One would like to show that they are pairwise inequivalent as $(\mathbf{g},K)$-modules, for any two distinct K-orbits Y_1 and Y_2. I cannot quite prove this statement, but a few remarks can be made. Recall that the algebra D_λ of global differential operators on X act on $M(Y_1,\lambda)$. Let I_λ be the annihilator in U of $M(Y_1,\lambda)$. I_λ is also the kernel of the map $U \to D_\lambda$. It is proven in [Borho-Brylinski I, Cor. 5.12] that D_λ and U/I_λ considered as U-bimodules not only have both $\mathcal{N}_p$ as associated variety but also that their quotient has an associated variety strictly smaller than $\mathcal{N}_p$. Hence the left and right annihilators in U/I_λ of D_λ are non-zero. By lemma 4.6, we conclude that D_λ has the same simple faithful modules as U/I_λ. Moreover, it is not hard to see that D_μ is isomorphic to the algebra of endomorphisms of $M(Y,\lambda)$ which are finite for the diagonal action of $\mathbf{g}$ and K.

If the two modules $M(Y_1,\lambda)$ and $M(Y_2,\mu)$ were equivalent as $(\mathbf{g},K)$-modules, then they would both have actions of D_λ and D_μ, and they would be equivalent as modules over these algebras. It follows that the D_λ and D_μ would be isomorphic. This is of course possible, but somewhat unlikely in the present context where both λ and μ are P-dominant. The condition that they be faithful modules, may not be important for the conjectural statement. (For example, it fails for compact groups, but the infinitesimal characters considered here are singular, *cf.* lemma 3.5, so that the corresponding representations of compact groups are zero anyway).

Consider the most singular case for $E_7/E_6+\mathbf{C}$: $\lambda=(4,1,1)$, $\mu=(1,3,1)$ and $\nu=(4,4,1)$. Then $s_7\,s_5\,s_6(\lambda_\rho)$ and $s_2\,s_1(\mu_\rho)$ are both equal to:

$$\begin{pmatrix} & & 1 & & & \\ 1 & 1 & 0 & 1 & 0 & 1 \end{pmatrix}$$

There are eight possibilities for the τ-invariant of the constituents of $A(\nu)$:

1. $\begin{pmatrix} & & - & & & \\ - & - & - & - & - & - \end{pmatrix}$
2. $\begin{pmatrix} & & - & & & \\ - & - & - & - & - & + \end{pmatrix}$
3. $\begin{pmatrix} & & - & & & \\ + & - & - & - & + & - \end{pmatrix}$
4. $\begin{pmatrix} & & - & & & \\ + & - & - & - & + & + \end{pmatrix}$
5. $\begin{pmatrix} & & - & & & \\ - & - & - & - & + & - \end{pmatrix}$
6. $\begin{pmatrix} & & - & & & \\ - & - & - & - & + & + \end{pmatrix}$
7. $\begin{pmatrix} & & - & & & \\ + & - & - & - & - & - \end{pmatrix}$
8. $\begin{pmatrix} & & - & & & \\ + & - & - & - & - & + \end{pmatrix}$

We start with a constituent M of $A(\nu)$ and cross the walls one by one until we reach the weights in question, so there is no ambiguity. In Case 3 going to λ_ρ we get zero, but going to μ_ρ we get $\begin{pmatrix} & & - & & & \\ - & - & + & - & + & - \end{pmatrix}$ as τ-invariant. In Case 4 going to λ_ρ we get $\begin{pmatrix} & & + & & & \\ - & + & - & + & - & - \end{pmatrix}$, and going to μ_ρ yields

$$\begin{pmatrix} & & - & & & \\ - & - & + & - & - & + \end{pmatrix}.$$

In Case 5 going to λ_ρ or μ_ρ yields zero. In Case 6 going to λ_ρ gives $\left(\begin{matrix} & & - & & & \\ - & - & + & - & + & - \end{matrix}\right.$ as τ-invariant and going to μ_ρ is zero. In Case 7 we get zero on the λ-side and $\begin{pmatrix} & & - & & & \\ - & - & + & - & - & - \end{pmatrix}$ on the μ-side. In Case 8 we obtain nothing on the λ-side and $\begin{pmatrix} & & - & & & \\ - & - & + & - & - & + \end{pmatrix}$ for μ_ρ. Thus ambiguities can arise only between μ_ρ in Case 3 and λ_ρ in Case 4 and λ_ρ in Case 8. The remarks we made at the end of the E_6/F_4 case, apply also here, if one wants to show that the $(\mathbf{g}, K)$-modules $M(Y_1, \lambda)$ and $M(Y_2, \mu)$ are inequivalent.

Finally, the case $E_8/E_7 + \mathbf{sl}(2)$ is completely analogous.

Bibliography

E. van den Ban 1: Invariant differential operators on a semisimple symmetric space and finite multiplicities in a Plancherel formula. Arkiv för Matematik **25** (1987) 175–187.

E. van den Ban 2: The principal series for a reductive symmetric space I: H-fixed distribution vectors. Ann. Sci. Éc. Norm. Sup. **21** (1988) 359–412.

E. van den Ban–P. Delorme: Quelques propriétés des représentations sphériques pour les espaces symétriques pour les espaces réductifs. J. Func. Anal. **80** (1988) 284–307.

H. Bass: *Algebraic K-theory*. Math. Lect. Notes Ser. Benjamin, New York-Amsterdam, 1968.

A. Beilinson: Localization of reductive Lie algebras. Proc. Int. Cong. Math. 699–710, Warszawa 1983.

A. Beilinson–J. Bernstein: Localisation de g-modules C.R.A.S. Paris **292** (1981) 15–18.

A. Beilinson–J. Bernstein: A generalization of Casselman's submodule theorem. In *Representation theory of reductive groups*. Ed. P. Trombi. Progress in Math. **40**, 35–52, Birkhäuser 1983.

Y. Benoist: Multiplicité un pour les espaces symétriques exponentiels. Mém. Soc. Math. France **15** (1984) 1–37.

J. Bernstein: Lecture notes (mimeographed), Harvard U., 1983.

W. Beynon–N. Spaltenstein: Green functions of finite Chevalley groups of type E_n. J. Algebra **88** (1984) 584–614.

F. Bien 1: Affine Weyl groups with involutions Canadian Math. Soc. Conf. Proc. vol. 6 (1986) 11–16.

F. Bien 2: Spherical $\mathcal{D}$-modules and representations of reductive Lie groups. Ph.D. dissertation, Massachusetts Institute of Technology, May 1986.

F. Bien 3: Spherical $\mathcal{D}$-modules and representations on symmetric spaces. Bull. Soc. Math. Belgique **38** (1986) 37–44.

A. Borel et al.: *Algebraic D-modules.* Perspectives in Math. **2**, Academic Press, Orlando 1987.

W. Borho–J.L. Brylinski: Differential operators on homogeneous spaces I. Invent. Math **69** (1982) 437–476. III. Invent. Math **80** (1985) 1–68.

W. Borho–R. MacPherson: Representations des groupes de Weyl et homologie d'intersection pour les variétés nilpotentes. C.R.A.S. Paris **292** (1981) 707–710.

R. Bott–L. Tu: *Differential forms in algebraic topology.* Grad. Text Math. vol. 82 Springer-Verlag 1982.

M. Brion: Quelques Propriétés des espaces homogènes sphériques. Manuscripta Math. **55** (1986) 191–198.

W. Casselman: Canonical extensions of Harish-Chandra modules to representations of G. Preprint, University of British Columbia, Vancouver.

C. De Concini–C. Procesi: Complete symmetric varieties. Lect. Notes Math. vol 996, Springer-Verlag 1983.

T. Enright–N. Wallach: Notes on homological algebra. Duke J. Math **47** (1980) 1–15.

M. Flensted–Jensen: Discrete series for semisimple symmetric spaces. Annals Math. **111** (1980) 253–311.

V. Ginzburg: Characteristic varieties and vanishing cycles. Invent. Math. **84** (1986) 327–402.

R. Godement: A theory of spherical functions I. Trans. A.M.S. **73** (1952) 496–566.

Harish-Chandra: Representations of semisimple Lie groups IV. Proc. N.A.S. **37** (1951) 691–694.

S. Helgason 1: Fundamental solutions of invariant differential operators on symmetric spaces. Amer. J. Math. **86** (1964) 565–601.

S. Helgason 2: *Differential geometry, Lie groups and symmetric spaces.* Pure and Appl. Math. vol. 80, Academic Press, 1978.

S. Helgason 3: *Groups, and Geometric Analysis* Pure and Appl. Math. vol. 113 , Academic Press, 1984.

S. Helgason 4: Some results on invariant differential operators on symmetric spaces. Preprint, Massachusetts Institute of Technology, May 1989.

W. Hesselink: Polarizations in the classical groups. Math. Z. **160** (1978) 217–234.

M. Kashiwara 1: *Seminar in microlocal analysis.* coauthors V. Guillemin and T. Kawai, Ann. Math. Studies **93**, Princeton Univ. Press 1979.

M. Kashiwara 2: The Rieman-Hilbert problem for holonomic systems. RIMS Publication **437**, Kyoto, 1983.

M. Kashiwara, A. Kowata, K. Minemura, K. Okamoto, T. Oshima and M. Tanaka: Eigenfunctions of invariant differential operators on a symmetric space. Ann. Math. **107** (1978) 1–39.

T, Kengmana: Characters of the discrete series for pseudo-Riemannian symmetric spaces. In *Representation theory of reductive groups.* Ed. P. Trombi. Progress in Math. **40**, 177–183, Birkhäuser 1983.

B. Kostant: On the existence and irreducibility of certain series of representations. Bull. A.M.S. **75** (1969) 627–642.

B. Kostant–S. Rallis: Orbits and representations associated with symmetric spaces. Amer. J. Math. **93** (1971) 753–809.

H.-P. Kraft–C. Procesi: On the geometry of conjugacy classes in the classical groups. Comment. Math. Helv. **57** (1982) 539–602.

T. Matsuki: The orbits of affine symmetric spaces under the action of minimal parabolic subgroups. J. Math. Soc. Japan **31** (1979) 331–357.

H. Matsumoto: Quelques remarques sur les groupes de Lie algébriques réels. J. Math. Soc. Japan **16** (1964) 419–446.

R. Richardson: Differential of the quotient morphism for a semisimple Lie algebra. In preparation, Austalian National University, Canberra.

T. Oshima–T. Matsuki: A description of discrete series for semisimple symmetric spaces. Advanced Studies in Pure Mathematics **4** (1984) 331–390.

P. Shapira: *Microdifferential systems in the complex domain.* Grundlehren **269**, Springer-Verlag 1985.

H. Schlichtkrull: *Hyperfunctions and harmonic analysis on symmetric spaces.* Progress Math. vol. 49, Birkhäuser, 1984.

T. Springer: Algebraic groups with involutions. Can. Math. Soc. Conf. Proc. vol. 6 (1986).

T. Springer–R. Steinberg: Conjugacy classes. Lecture Notes Math. vol. 131, 167–266; Springer-Verlag, 1970.

R. Steinberg: On the desingularization of the unipotent variety. Inv. Math. **36** (1976) 209–224.

F. Treves: *Topological vector spaces, distributions and kernels* Pure & Applied Math. **25** Academic Press, New York-London 1967.

D. Vogan 1: A generalized τ-invariant for the primitive spectrum of a semisimple Lie algebra. Math. Ann. **242** (1979) 209.

D. Vogan 2: *Representations of real reductive Lie groups.* Progress Math. **15**, Birkhäuser, Boston-Basel-Stuttgart 1981.

D. Vogan 3: Unitarizability of series of representations. Annals Math. **120** (1984) 141–187.

D. Vogan 4: The orbit method and primitive ideals for semisimple Lie algebras. In *Lie algebras and related topics.* Britten, Lemire and Moody eds. Proc. Can. Math. Soc. Conf. **5** (1986).

D. Vogan 5: Irreducibility of discrete series representations for semisimple symmetric spaces. In *Representations of Lie Groups*, Kyoto-Hiroshima, Advanced Studies in Pure Math. **14**, 1988.

N. Wallach: Asymptotic expansions of generalized matrix entries of representations of real reductive groups. Lect. Notes Math. **1024**, 287–369, Springer-Verlag 1983.

G. Zuckerman: Geometric methods in representation theory. In *Representation theory of reductive groups.* Ed. P. Trombi, Progress in Math. **40**, 283–290. Birkhäuser 1983.

Index

$A_K(G_o/H_o, \chi)$ 36
ambiguous infinitesimal
  character 93
admissible subgroup 26
affine imbedding 29
affinity of flag spaces 20
associated variety of
  a $(\mathbf{g}, K)$-module 77

$\mathcal{B}$ as flag space 29
$\mathcal{B}$ as hyperfunctions 52
$B(X^r, L_\lambda)$ 37, 52
Blattner formula 62

C_p 74, 84
Cartan factor 12
Cartan triple 13
characteristic cycle 83
characteristic variety 69
classification of
  $(\mathcal{D}, K)$-modules 26
closed orbits 32
coadjoint orbits 117
coherent family 119
compactification of a
  symmetric space 63, 66
component group 50

Δ_λ 20,22
$\mathcal{D}_\lambda$ 15
$\mathcal{D}_\ell$ 16
$\mathcal{D}_{\mathbf{t}_p}$ 16
$\mathcal{D}$-module 9
  coherent – 24
  holonomic – 25
  irreducible – 25
  smooth – 24
  standard – 25, 27
  regular singularities – 25
$(\mathcal{D}, G)$-module 11
decomposable module 80
discrete series 58
  construction of – 40
duality of functors 40

$\mathcal{E}_X$, $\mathcal{E}_\lambda$ 68
$\mathcal{E}$-module 68
  holonomic – 69
exactness of Γ 20, 23

flag space 12, 29
Flensted-Jensen
  isomorphism 37

G, G_o 1
G^r 2
Γ as functor of
  global sections 20
Γ as functor of
  K-finite sections 40

H, H_o 1
H^τ 2
Harish-Chandra isomorphism 15
harmonic polynomial 16
Helgason isomorphism 37
hyperfunctions 37, 52

inductive functor 39
invariant differential operators 35, 91
irreducibility criterion 78

K, K_o 1
K^τ 2

λ_ρ 119
Levi factor 12
localization functor 20
L as functor of cofinite sections 40

μ_ρ 119
matrix differential operator 11, 105
mic as functor of microlocalization 72
microdifferential operator 68
moment map 74
multiplicity one 61, 97, 108
multiplicities 4, 116

$\mathcal{N}_\mathfrak{p}$ 74, 87
nilpotent cone 85
nilpotent variety 85

$\mathcal{P}$ 29
parabolic subgroup 12
Poisson transform 37, 64
positive roots 13
primitive ring 107
projective functor 39

ρ, $\rho_\mathfrak{b}$, $\rho_\mathfrak{p}$, $\rho_\mathfrak{l}$ 13
rank 58
real forms of an algebraic group 34
Richardson orbit 74
Riemannian dual of a symmetric space 2
root system 13

seminormal point 86
spherical module 34, 45
Springer correspondence 87

τ-invariant 118
Thom isomorphism 53
trivial isotropy 33, 45

U_χ, $\mathcal{U}_\chi$ 73
$U_\mathfrak{l}$ 16
unibranch point 86
unitary module 82

X as flag space 12

Z, $Z(\mathfrak{g})$ as center of U 15